T0185836

A Journey of Embedded and Cyber-Physical Systems

Jian-Jia Chen

Editor

A Journey of Embedded and Cyber-Physical Systems

Essays Dedicated to Peter Marwedel
on the Occasion of His 70th Birthday

 Springer

Editor
Jian-Jia Chen
Dortmund, Germany

ISBN 978-3-030-47489-8   ISBN 978-3-030-47487-4   (eBook)
https://doi.org/10.1007/978-3-030-47487-4

This Springer imprint is published by the registered company Springer Nature Switzerland AG.
The registered company address is: Gewerbestrasse 11, 6330 Cham, Switzerland

Dedicated to Prof. Peter Marwedel

Foreword

Professor Dr. Peter Marwedel has made many invaluable contributions to research in the Electronic Design Automation area and specifically to the field of Embedded System Design. Already in his young years, he placed a landmark in the field of high-level synthesis and hardware description languages. In the mid-1970s, he started to work on high-level synthesis—almost a decade before it became mainstream. Peter Marwedel also laid ground for the field of retargetable compilation, which he considered as a special case of high-level synthesis in which the architecture was fixed. He designed the very first approach for generating a compiler from a description of the processor architecture and published this work already in the early 1980s—long before compilers were an accepted topic at EDA conferences. In the early 1990s, Peter Marwedel realized that compilers for embedded processors will become a crucial element in the toolchain for designing efficient embedded systems. One of his most prominent activities in the area of compilers for embedded systems is that of energy-aware compilation where he considered the energy efficiency of compiled code. His work on exploiting scratchpad memories for improving energy efficiency is well-known throughout the world. His engagement in memory-architecture aware compilation is highly influential and can be seen as an early contribution towards green computing.

In addition to being an outstanding researcher, Peter Marwedel has also proven himself as a great educator and passionate teacher in the classroom. This is best evidenced by his textbook "Embedded System Design" which has become standard literature in higher-level education on Embedded Systems for over a decade now. His excellence in teaching is documented, e.g., by the invitation of the Advanced Institute of Information Technology (AIIT) to give a 1-week compact course on embedded system design to Korean professors in Seoul in 2008 or by an invited talk on embedded system education during the cyber-physical systems program at the annual meeting of the US National Science Foundation (NSF).

While this Festschrift features a variety of contributions from his peers in the professional community, we would like to take this opportunity to emphasize Peter Marwedel's role as an exemplary advisor for his students, all the way from freshmen studying for their BS and MS degrees to PhD graduates and rising university professors. His dedicated mentorship and valuable advice has had a tremendous impact on many careers, as the following personal quotes show:

> Back in July 1995, I was a student at the University of Dortmund, working hard to complete my diploma thesis. At that time, my advisor Prof. Marwedel left Dortmund for a sabbatical at the University of California. I vaguely remember the day when he sent me an email, asking me if I would be interested to study for PhD in the USA. It took me less than 5 min to respond with 'No, thank you!' as I had always envisioned to spend my life and work in Germany. Nine days later Peter Marwedel changed my life with a second email that essentially read 'Think twice!'. So I did. Long story short, I moved to California and live there since. I will be forever grateful for that second email which opened up a once-in-a-lifetime opportunity for my successful career. Thank you!
>
> Rainer Dömer (Professor, University of California at Irvine, USA)

> I got in touch with Peter Marwedel's chair for the first time in 1996 in the context of my student project work on HW/SW co-design and FPGA synthesis. With my successful graduation 1998, Peter offered me a PhD position in his group so that I had the chance to collaborate with the IMEC in Leuven on high-level source code optimization techniques. Having completed my PhD in 2004, Peter continued to support me so that I was able to establish a team working on compilers for real-time systems. Besides this excellent mentorship for more than a decade that finally led my to my current professorship at TUHH, I always appreciated Peter Marwedel's team spirit very much. Besides professional activities, team building was always very important for him, leading to countless memories of garden parties, barbecues with students in his chair's backyard, bicycle tours, workshops in the middle Rhine Valley, etc. This high degree of support and friendship is extraordinary and I am happy about the opportunity to say thank you to Peter Marwedel in this Festschrift.
>
> Heiko Falk (Professor, Hamburg University of Technology, Germany)

> Even though I started as a postdoc in Peter Marwedel's research group in 2010, I was able to learn a lot from his decades of experience in teaching, scientific writing, and designing research proposals. We were always eager to experiment with new approaches, such as flipped classroom teaching, and to explore new research topics. All of this has had a significant influence on my current work as professor at Coburg University. The extensive research on compilers in Peter's group also contributed to obtaining my new position at NTNU. An invaluable benefit when working in a large, well-funded research group was that I got to travel to so many interesting places and meet colleagues from all around the world. Traveling was also a common topic on the private level, since Peter loves to travel and to photograph. I fondly remember many hours of browsing through Peter's diligently prepared photo albums and listening to the stories of his trips. Going to Norway, I hope that I can provide an exciting destination for an upcoming trip. Thanks a lot for everything, Peter, and see you soon in Trondheim!
>
> Michael Engel (Professor, NTNU, Trondheim, Norway).

Numerous anecdotes just like these could be added here, but that would fill an entire book by itself. So without further ado, we conclude this foreword here with a big THANK YOU to Peter Marwedel.

Workshop at TU Dortmund, July 2019   Rainer Dömer
                                                                                                               Heiko Falk
                                                                                                               Michael Engel

Acknowledgements

The workshop on embedded systems, dedicated to Peter Marwedel on the occasion of his 70th birthday, and its festschrift have been partially supported by Deutsche Forschungsgemeinschaft (DFG), Collaborative Research Center SFB 876 (http://sfb876.tu-dortmund.de/), and Alumni der Informatik e.V. TU Dortmund.

Contents

Contributors

Hussam Amrouch Karlsruhe Institute of Technology (KIT), Karlsruhe, Germany

Emad Arasteh Center for Embedded and Cyber-Physical Systems, University of California, Irvine, CA, USA

M. Balakrishnan Indian Institute of Technology Delhi, Delhi, India

Jian-Jia Chen TU Dortmund, Dortmund, Germany

Kuan-Hsun Chen TU Dortmund, Dortmund, Germany

Zhongqi Cheng Center for Embedded and Cyber-Physical Systems, University of California, Irvine, CA, USA

Bryan Donyanavard University of California, Irvine, CA, USA

Rainer Dömer Center for Embedded and Cyber-Physical Systems, University of California, Irvine, CA, USA

Caio Batista de Melo University of California, Irvine, CA, USA

Nikil Dutt University of California, Irvine, CA, USA

Heiko Falk Institute of Embedded Systems, Hamburg University of Technology (TUHH), Hamburg, Germany

Gernot Fink TU Dortmund, Dortmund, Germany

Gernot Gebhard AbsInt Angewandte Informatik GmbH, Saarbrücken, Germany

Jörg Henkel Karlsruhe Institute of Technology (KIT), Karlsruhe, Germany

Wen-Hung Huang TU Dortmund, Dortmund, Germany

Shashank Jadhav Institute of Embedded Systems, Hamburg University of Technology (TUHH), Hamburg, Germany

Matthias Jung Fraunhofer IESE, Kaiserslautern, Germany

Daniel Kästner AbsInt Angewandte Informatik GmbH, Saarbrücken, Germany

Arno Luppold Institute of Embedded Systems, Hamburg University of Technology (TUHH), Hamburg, Germany

Pouya Mahmoody Friedrich-Alexander-Universität Erlangen-Nürnberg, Erlangen, Germany

Biswadip Maity University of California, Irvine, CA, USA

Daniel Mendoza Center for Embedded and Cyber-Physical Systems, University of California, Irvine, CA, USA

Kasra Moazzemi University of California, Irvine, CA, USA

Kateryna Muts Institute of Embedded Systems, Hamburg University of Technology (TUHH), Hamburg, Germany

Tiago Mück University of California, Irvine, CA, USA

Heinrich Müller TU Dortmund, Dortmund, Germany

Dominic Oehlert Institute of Embedded Systems, Hamburg University of Technology (TUHH), Hamburg, Germany

Nina Piontek Institute of Embedded Systems, Hamburg University of Technology (TUHH), Hamburg, Germany

Markus Pister AbsInt Angewandte Informatik GmbH, Saarbrücken, Germany

Behnaz Pourmohseni Friedrich-Alexander-Universität Erlangen-Nürnberg, Erlangen, Germany

Martin Rapp Karlsruhe Institute of Technology (KIT), Karlsruhe, Germany

Amir M. Rahmani Technische Universität Wien, Vienna, Austria

Sascha Roloff Friedrich-Alexander-Universität Erlangen-Nürnberg, Erlangen, Germany

Mikko Roth Institute of Embedded Systems, Hamburg University of Technology (TUHH), Hamburg, Germany

Sami Salamin Karlsruhe Institute of Technology (KIT), Karlsruhe, Germany

Wolfgang Schröder-Preikschat Friedrich-Alexander-Universität Erlangen-Nürnberg, Erlangen, Germany

Kenneth Stewart University of California, Irvine, CA, USA

Jürgen Teich Friedrich-Alexander-Universität Erlangen-Nürnberg, Erlangen, Germany

Niklas Ueter TU Dortmund, Dortmund, Germany

Georg von der Brüggen TU Dortmund, Dortmund, Germany

Norbert Wehn TU Kaiserslautern, Kaiserslautern, Germany

Christian Weis TU Kaiserslautern, Kaiserslautern, Germany

Stefan Wildermann Friedrich-Alexander-Universität Erlangen-Nürnberg, Erlangen, Germany

Reinhard Wilhelm Universität des Saarbrücken, Saarbrücken, Germany

Saehanseul Yi University of California, Irvine, CA, USA

Chapter 1
Peter Marwedel and the Department of Computer Science of the TU Dortmund University

Gernot Fink and Heinrich Müller

1.1 Introduction

Peter Marwedel was appointed Professor at the Department of Computer Science of the University of Dortmund in 1989. He represented the area "Computer Engineering and Embedded Systems" and headed the Chair of Computer Science 12 until his retirement in 2014. During this time, he has made a great contribution to the Department of Computer Science, which continues to have a lasting effect today. The following presentation is intended to give an idea of the extraordinary breadth and importance of his activities in teaching, academic self-government, basic research, and technology transfer.

1.2 Teaching

A special passion of Professor Marwedel is teaching, which is not self-evident for a dedicated and successful researcher. In Dortmund, he was extremely involved in basic teaching, which is particularly challenging given the high number of students attending these courses. For many years he has held the introductory compulsory lecture "Computer Structures" and the elective lectures "Embedded Systems" and "Computer Architecture." Through these courses, he has substantially contributed to the basic training in technical computer science. His lectures were extremely well prepared. The material was thoughtfully selected in all details, didactically carefully prepared, and presented objectively and clearly. The courses have gained a high

G. Fink · H. Müller (✉)
TU Dortmund, Dortmund, Germany
e-mail: Gernot.Fink@tu-dortmund.de; heinrich.mueller@cs.uni-dortmund.de

© The Author(s) 2021

J.-J. Chen (ed.), *A Journey of Embedded and Cyber-Physical Systems*,
https://doi.org/10.1007/978-3-030-47487-4_1

1

reputation among the students. This was reflected in consistently excellent student ratings and led to Professor Marwedel receiving the prestigious Teaching Award of TU Dortmund University in 2003.

From the course "Embedded Systems" the first edition of the English textbook "Embedded System Design" emerged in 2003. The book has established itself as an international textbook and is often cited. Professor Marwedel has adapted it over the years to current developments. The third edition has been published in 2018.

In addition to traditional teaching, Professor Marwedel has shown great interest in the possibilities of new media and has made strong use of it. On YouTube he has made available a significant number of educational videos that are particularly associated with his courses "computer structures" and "computer architecture." The design of the videos focuses on the essentials and avoids superfluous, distracting visual effects.

In his last active years at the department, Professor Marwedel experimented with the concept of the inverted classroom as a natural consequence. This form of teaching is still little used at German universities.

Particularly noteworthy is the fact that Professor Marwedel has conveyed his concept of education in the field of embedded systems beyond its actual implementation. It was the subject of the Workshop on Embedded and Cyber-Physical System Education (WESE) organized by him in Finland in 2012. In the same year, he was invited to the annual meeting of the cyber-physical systems program of the prestigious National Science Foundation (NSF) in Washington, USA, where he gave a talk focusing on the Dortmund education concept in the field of Embedded Systems.

Finally, it should be mentioned that Professor Marwedel contributed to the internationalized teaching of the Department by participating in the English Master's program "Automation and Robotics" of the TU Dortmund University and his contacts to Indian Institutes of Technology, which have led to internships of Indian students in Dortmund.

1.3   Academic Self-Government

Professor Marwedel has always been active in the Department's academic administration and beyond. He was Dean of Studies of the Department and member of the Academic Senate of TU Dortmund. But here, too, his commitment to teaching is particularly reflected. He was chairman of the teaching committee of the Department and chairman of the educational committee of the Academic Senate of the university. He was Dean of Studies of the Department from 2012 to 2014. Beyond the university, he has worked on standards for the accreditation of computer science curricula in the national organizations ASIIN and AVI.

1.4 Basic Research and SFB 876

Professor Marwedel has contributed significantly to the international visibility of the Department of Computer Science through his research in embedded and cyber-physical systems. He has received several international honors, in particular the EDAA Lifetime Achievement Award (2013), the ESWEEK Lifetime Achievement Award (2014), and the ACM SIGDA Distinguished Service Award (2014). In 2010 he became an IEEE Fellow and a Fellow of the Design, Automation and Test Conference in Europe (DATE).

A very special contribution to research at the Department and the university was made by Professor Marwedel as co-initiator of the Collaborative Research Center (CRC/SFB) 876, together with the main initiator Prof. Dr. Katharina Morik, spokeswoman for the CRC. The CRC program of the Deutsche Forschungsgemein-schaft (DFG) enjoys a high reputation and is extremely competitive. The topic of the SFB 876 is "Providing Information by Resource-Constrained Data Analysis." The collaborative research center SFB 876 brings together data mining and embedded systems. On the one hand, embedded systems can be improved using machine learning. On the other hand, data mining algorithms can be realized in hardware, e.g. FPGAs, or run on GPGPUs. The restrictions of ubiquitous systems in computing power, memory, and energy demand new algorithms for known learning tasks. At the time of the application, merging data analysis and resource constraints was visionary—today it is highly relevant in many applications. In the SFB 876, about 20 research groups of the Department, of TU Dortmund University, and of neighboring universities are working together since 2011 for 12 years in an interdisciplinary manner.

1.5 Technology Transfer and ICD

Professor Marwedel has not only been active in basic research, but has also dealt with applied research and the transfer of scientific results to the economy. For many years he is CEO of the "Informatik Centrum Dortmund e.V." (ICD) and head of the Embedded Systems Group at ICD. The ICD was founded in 1989 from the Department of Computer Science at the University of Dortmund. The ICD is available to companies from all sectors of the economy. Its goal is to accelerate the transfer of current research results in computer science and information technology into industrial products. Under Professor Marwedel's leadership, the ICD has become a well-established, successful, and economically stable association.

1.6 Conclusion

The Department of Computer Science is extremely grateful to Professor Marwedel. On the occasion of his seventieth birthday, it wishes him all the best for the future.

Chapter 2
Testing Implementation Soundness
of a WCET Analysis Tool

Reinhard Wilhelm, Markus Pister, Gernot Gebhard, and Daniel Kästner

2.1 Introduction

Timing verification of a set of hard real-time tasks to be executed on a given hardware platform attempts to prove that all tasks in the set when executed on that platform always respect their deadlines, i.e., each task finishes its execution within its deadline. Traditionally, timing verification is split into two subtasks: a *timing analysis* also known as *WCET analysis*, which statically determines upper bounds on the execution times of the tasks, and a *schedulability analysis*, which takes these upper bounds and attempts to verify that all tasks in the given set, assuming these upper bounds on their execution times, will respect their deadlines.

A preliminary version of this paper appeared in [16].

R. Wilhelm (✉)
Universität des Saarbrücken, Saarbrücken, Germany
e-mail: wilhelm@cs.uni-saarland.de

M. Pister · G. Gebhard · D. Kästner
AbsInt Angewandte Informatik GmbH, Saarbrücken, Germany
e-mail: pister@absint.com; gebhard@absint.com; kaestner@absint.com

J.-J. Chen (ed.), *A Journey of Embedded and Cyber-Physical Systems*,
https://doi.org/10.1007/978-3-030-47487-4_2

2.1.1   Tool Qualification

WCET analysis is applied to time-critical and safety-critical embedded-system software in problem-aware parts of the embedded-systems industry. Such systems have to be developed in accordance with international safety norms, e.g., DO-178B/C, DO-254, IEC 61508, and ISO 26262. While there are differences between these norms, in particular regarding prescriptiveness and required level of rigor, they have many aspects in common. All of them include guidance on the use of software tools as a part of the development and verification process of safety-critical software.

The criticality level (also known as the design assurance level (DAL) or safety integrity level (SIL)) of a component determines the effort to invest and the methods required or recommended to deliver assurance of the correct functioning of the component. The criticality level is derived from the impact of a failure of the component on the functioning of the system. Similarly, the required activities to provide confidence in the correct functioning of a software tool depend on its criticality with respect to the overall system. For example, DO-178C, the current international standard for avionics systems, defines five different *tool qualification levels (TQLs)*. The TQL is determined by the potential tool impact and the design assurance level of the software. There are three tool-impact categories; the most critical, *Category 1*, applies to tools whose output becomes part of the airborne software. Similar considerations are also made in other norms, e.g., the ISO 26262 defines a *tool confidence level (TCL)* in a very similar way.

The overall goal of *tool qualification* is to provide confidence that the tool operates correctly, i.e., according to its functional specification, in the operational context of the tool user. In the following, we will focus on the tool qualification requirements of the avionics industry, which are the most rigid of the safety-critical industries. Certification of avionics systems is regulated by the international standard DO-178C [1]. WCET analysis tools fare under *verification tools*. Verification tools have no overly rigid certification requirements, unlike *development tools*: their impact category is *Category 2* or *Category 3*, mostly depending on whether the output of the tool is used to justify the elimination or reduction of other verification or development activities or not. A prerequisite for tool qualification is a specification of the tool functionality. The *tool operational requirements (TOR)* specify the tool functions and technical features, which are stated as low-level requirements on tool behavior under normal operating conditions. Another required input is the *verification test plan (VTP)*, which defines test cases demonstrating the correct functioning of all specified requirements of the TOR. Test-case definitions include the overall test setup as well as a detailed structural and functional description of each test case, i.e., how the individual test case works and what the expected result is.

Certification becomes more challenging through DO-333, the formal-methods supplement to DO-178C. It asks for a statement that a formal method including the underlying theory is *adequate* for solving the corresponding verification problem. This introduces and enforces *soundness* of the methods and tools.

Since the required effort for tool qualification can be high, ideally the software qualification process is supported by a *qualification support kit (QSK)* supplied by the tool provider. It must include TOR and VTP and typically provides a validation suite, which allows users to execute the relevant test cases in the relevant operational context. TOR, VTP, and a test execution report become part of the certification package. Furthermore, it is typically required for a tool provider to supply *qualification software life cycle data* to demonstrate that the development process and the invested efforts to assure correctness, quality, and traceability are adequate for usage in a safety-critical system context. The qualification software life cycle is not covered in this article.

DO-178C exhales a test-based spirit: many verification activities are test based. Well-defined coverage criteria try to capture to which extent the behavior of the system under test has actually been exerted during testing. Note that in case of a static verification tool, test coverage does not apply to the code to be analyzed: a sound static-analysis tool provides full data and control coverage, i.e., it analyses all paths and takes into account all potential data values for its analyses. What is needed in case of the microarchitectural analysis, which is the focus of this article, is to demonstrate the correctness of the microarchitecture model used by the analyzer. To this end it is the instruction set architecture (ISA) and the set of paths through the execution platform that need to be covered. Huge sets of test traces in qualification suites are used at tool-qualification time to cover the sets of paths through the execution platform.

Note the difference to measurement-based WCET analyses. It is known that they are in general unsound. In order to provide a sufficient level of confidence in the real-time behavior of industrial-size code they need an unacceptably huge set of traces and accordingly an excessive effort at verification time. In the case of a static WCET analysis tool, the testing effort is applied at tool-qualification time when ample time is available.

2.1.2   Predictability

Timing predictability [3, 15] has long been recognized as essential for achieving precise results of timing estimation at reduced analysis effort. In the context of the current article, it is worth mentioning that it also reduces the number of test cases for the validation of an abstract architectural model. In general, an increase in the timing predictability of the underlying architecture leads to a decreasing number of different instruction flow paths through the processor pipeline since they feature less average-case performance-enhancing micro-optimizations like instruction and data queues and buffers, data forwards, etc. Such architectures show a more regular hardware design.

2.1.3   WCET Analysis

Performance-enhancing architectural components such as caches, pipelines, and speculation have made WCET analysis difficult. Execution times of consecutively executed instructions do not compose easily because instruction execution times are now dependent on the execution state in which they are executed. In the composition *A;B* the execution time of statement *B* depends on the execution state produced by executing statement *A*. The variability of execution times grows with several architectural parameters, e.g., the cache-miss penalty and the costs for pipeline stalls and for control-flow mispredictions. As approaches using exhaustive measurements are infeasible due to the size of the search space, abstraction is applied leading to an over-approximation of the set of potential executions. This over-approximation introduces remaining uncertainty in the results of the microarchitectural analysis, which grows with the same architectural parameters mentioned above unless the architectural platform is predictable [18], see Sect. 2.1.2.

2.1.4   The Central Idea: Proving Safety Properties

We needed to solve the WCET problem for architectures with state-dependent execution times. Figure 2.1 shows that this problem could be decomposed into many subproblems. The main problem, specific for WCET analysis, was the *microarchitectural analysis*, a combined cache and pipeline analysis. Let us describe the central idea behind this phase in our WCET analysis method [17], first in a conceptual way, i.e., not quite like it is implemented, later closer to how it is implemented:

- We define any architectural effect that causes an instruction to execute longer than its fastest execution time to be a *timing accident*. Typical such timing accidents are cache misses, pipeline stalls, bus-access conflicts, or branch mispredictions. Each timing accident is associated with a *timing penalty*. Timing penalties may be constant, but may also be execution-state dependent. A cache-miss penalty may be constant if the bus is always guaranteed to be free for the cache reload. If this guarantee cannot be given, however, its size depends on the execution state, namely whether the bus happens to be free.

  The property that the execution of an instruction at some program point will not cause a particular timing accident is then a safety property. The occurrence of a timing accident thus violates a corresponding safety property.
- We then use an appropriate method for the verification of safety properties to prove that for the instructions in the program some of the potential timing accidents will never happen. The goal is to prove as many of such safety properties as possible. Conceptually, the safety properties shown to hold could be used to reduce the worst-case execution-time bound for an instruction, which a naive, sound WCET analysis would have to assume, by the cost for the excluded

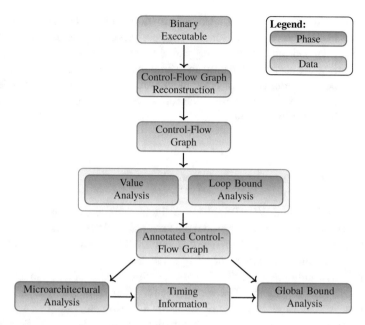

Fig. 2.1 The architecture of the aiT tool

timing accidents. In practice, pipeline analysis drives a cycle-wise transition, which considers the abstract execution state, e.g., makes no transition under a cache miss if a cache miss can be excluded.

- We then prove these safety properties by abstract interpretation (AI) [4] in the following way: Compute invariants at each program point, in our case an over-approximation of the set of execution states that are possible when execution reaches this program point. Derive the above mentioned safety properties, that certain timing accidents will not happen, from these invariants. For example, AI computes an abstract cache state at each program point, which overapproximates the sets of concrete cache states that may reach this program point. The abstract cache states are used to classify some memory accesses as definite hits. Another cache analysis that underapproximates the set of possible concrete cache states is able to predict definite misses. Predicted cache hits are then used to prove that the timing accident, this memory access will miss the cache, will never happen [8, 10].

  This method for the microarchitectural analysis was the main innovation that made our WCET analysis work for real-life architectures and scale to industrial-size software [6].

Now follows the description of the microarchitectural analysis that is closer to the implementation. Driver of this analysis is the pipeline analysis [14]. It goes through the instruction stream, instruction by instruction, and executes the current instruction in the current abstract execution state. This abstract execution state contains uncertainty, i.e., it lacks information about some state components. Transitions to all potential successor states are performed whenever the transition to the next state depends on such a missing part of the state. The timing contributions of these transitions are accumulated until an instruction can be retired. In the end, upper bounds on the execution times of basic blocks are obtained that are coefficients in an integer linear program representing the control flow of the program [17]. Another type of result is described below.

2.1.5   Terminology

We consider only sound WCET analysis methods. *Soundness* means that a method and associated tool will always produce *conservative* WCET estimates, i.e., estimates that will never be exceeded in any execution. Being conservative is a Boolean property. Unfortunately, *conservative* is often used as a metric property, *more conservative* meaning *less precise*. However, calling results of an unsound method conservative is a misnomer. The really meant, other dimension, in addition to soundness, is *accuracy*. Accuracy of some WCET estimate, obtained by a sound method, expresses the degree of over-estimation, the difference between a WCET estimate and the real WCET. It does not make sense to talk about the accuracy of an unsafe estimate or an unsound method. In case of an unsound method it is not even clear whether a "more conservative" estimate moves towards the real WCET from below or is larger than the real WCET and moves further away from it. In general, WCET estimates are below, i.e., underestimate the real WCET, if end-to-end measurements are used. On the other hand, if piecewise measurements are applied whose results are combined to an estimate of the overall execution times, this often results in over-estimation of the real WCET.

WCET analysis can be seen as the search for a longest path in the state space spanned by the program under analysis and by the architectural platform. Most real-time software is written as to guarantee termination. Its state space can thus be easily abstracted to a finite abstract state space, which is still too large to be exhaustively explored. We can, therefore, not expect to identify the real WCET, but only safe upper bounds to all execution times, which we will call WCET estimates. (Safe) over-approximation is used in several places. In particular, an abstraction of the execution platform is employed by the WCET analysis. How to convince oneself (or the certification authorities) of the correctness of this architectural model is the main subject of the next section.

2.2   Validation of Our WCET Analysis Tool

The claim that our WCET analysis tools produce safe results is a strong one and often disputed by some proponents of unsound WCET analysis methods. Their argument is, to develop an error-free instantiation of the, in principle, sound WCET analysis technology is so difficult, that one might use a simpler unsound method in the first place. The main complaint is the complexity of the abstract architectural models. So, what is the basis for our claims?

Several analyses in the tools are instances of abstract interpretation [4], a scientific method with a strong underlying theory, relating analysis results to semantic properties of analyzed programs. Value and loop bound analysis, c.f. Fig. 2.1, are more or less standard abstract interpretations. The difference is that these analyses are performed on the binary level and not on the source level. Still, adequacy of these analyses is easily accepted. The instantiation of the abstract-interpretation framework for the microarchitectural analysis of a given execution platform, however, is far from trivial. In particular, it contains an abstraction of the execution platform. How does one make sure that such an abstraction is conservative? This will be explained in Sect. 2.2.3.

Let us give short descriptions of the different component analyses alongside the particular validation activities before we come to the validation of the central component.

2.2.1 Control-Flow Graph Reconstruction

The reconstruction of the control-flow graph (CFG) from a binary executable means to compute a safe approximation of the inter-procedural control flow of the executable [13]. This is achieved by the following two steps after having loaded the executable:

1. Classification of the loaded byte stream to identify individual assembly instructions and
2. Recursive reconstruction of the control flow based on this assembly-instruction classifications.

For Step 1, a specification of the instruction encoding is required. Instruction-set-architecture manuals provide this information, which is then used to implement instruction identification in the binary decoder of the aiT tool chain. To validate the implementation, we perform the so-called *decode* tests. For each supported instruction (in each supported addressing mode) we write a test case providing a reference as the expected result of the decoding. The decoding result is then compared to this reference.

In Step 2, the decoder uses the identified instruction stream to compose a safe control-flow approximation. To validate this, we compile a representative set of control structures (in a high-level language like C) and decode the resulting executable to compare the reconstructed control flow with a reference result.

2.2.2   Value Analysis

The value analysis determines safe approximations of the values in processor registers and memory cells for every program point. These approximations are used to determine bounds on the iteration number of loops and information about the addresses of memory accesses. The value analysis is based on the instruction semantics of the underlying target architecture. Like the instruction encoding, architecture manuals provide this information.

To validate the instruction-semantics implementation, we create a test case for each instruction and define pre- and post-conditions according to the expected effect of the particular instruction. These conditions are expressed by user annotations, which are read by the value analyzer. Pre-conditions are used to generate the machine state needed to execute the tested instruction. The post-conditions define the expected state after having executed the instruction under test.

2.2.3   Microarchitectural Analysis: Trace Validation

The microarchitectural analysis combines a cache and a pipeline analysis. It is an abstract interpretation of the program's execution on the underlying cache and pipeline architecture. The execution of a program is abstractly executed by feeding instruction sequences from the control-flow graph to the timing model, which then computes the changes of the abstract execution state at cycle granularity and keeps track of the elapsing clock cycles. The correctness proofs of the method have been conducted by Thesing [14] based on the theory of abstract interpretation.

The cache analysis described in [2, 5, 7] is incorporated into the pipeline analysis. At each memory access, where the concrete hardware would query and update the contents of the cache(s), the cache analysis applies the corresponding abstract cache effects to the abstract cache state.

The result of the microarchitectural analysis is either an upper execution-time bound for every basic block or a *prediction graph*. In the first case, these upper bounds are the coefficients in an integer linear program that represents the control flow of the program. This is the version usually described in publications about static WCET analysis, as it presents a clean work distribution. However, it has the disadvantage that too much information is lost at basic-block boundaries, namely the precise matching of final states at predecessor blocks to initial states at successor blocks. This loss of information entails a loss in precision. The prediction graph avoids this loss of precision. It consists of abstract states as nodes and edges for the transition between states and represents the evolution of the abstract execution states at processor-clock granularity and beyond basic-block boundaries. Note that in the description of trace validation a prediction event graph appears, which is the prediction graph extended by event annotations at its edges.

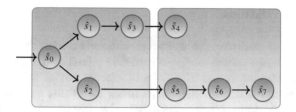

Fig. 2.2 Evolution of abstract hardware states $\hat{s}_i$. Each edge denotes a single cycle transition in the abstract state space. The gray boxes span the set of states that belong to the same basic block

As an example consider the prediction graph of Fig. 2.2, where the longest path is four transitions long, i.e., it takes four processor cycles to complete the program. Adding up the length of each longest path per basic block (denoted by the gray boxes) would neglect that there is no connection between the abstract states $\hat{s}_3$ and $\hat{s}_5$ and thus yield a worst-case estimate of five processor cycles.

Due to the complexity of the abstract architectural model, validation of the pipeline analysis cannot be done solely by testing the abstract implementation of individual instructions as we do it for CFG reconstruction and value analysis.

2.2.3.1   Semi-Automatic Derivation of the Abstract Architecture Model

Nowadays, hardware circuits are automatically synthesized from formal hardware specifications like VHDL or Verilog. Besides a formalization of the functional details, such specifications implicitly contain an execution model that also reflects the timing behavior of the whole system. It was a tempting idea to derive a pipeline analysis from the formal hardware model such that analysis and synthesized circuit share the same basis [11, 12].

However, the semi-automatic derivation of a timing model approach has not proven effective in the industrial context. Even if the hardware manufacturers grant access to their formal models (which is often not the case), the derivation process requires to fully understand the design, which might be a complex task for a complete processor including peripheral devices. Additionally, the quality of the resulting analysis depends on the coding style of the hardware model [11]. Results are excellent if the code features minimal dependencies between processes, a clear logical separation of different functionality into different processes/subprograms and a sequential logic design. Ideally, the code reflects the structural composition of the processor pipeline with explicit control signals to steer the flow of instructions and data. Models not adhering to those design principles complicate state abstractions and thus result in prohibitively resource-consuming analyses.

2.2.3.2 Trace Validation

For the reasons given above, the abstract architectural models are *hand-crafted* by human experts based on the available hardware reference documentation, which sometimes contains errors and usually lacks relevant details. Reverse engineering based on specific runtime measurements needs to fill this gap. Even if it were semi-automatically derived from a specification, the implementation of the microarchitectural analysis would still need to be validated. *Trace validation* checks for safe over-approximations of the predictions by matching observable hardware events recorded during concrete executions of instruction sequences against predictions of those events produced by the microarchitectural analysis. This is done for a sufficiently large set of instruction sequences that structurally covers the possible instruction flows (wrt. the different functional units, instructions, dependencies between instructions, etc.) of the processor pipeline.

Figure 2.3 shows the trace-validation workflow. An instruction sequence is executed on the actual hardware, or its execution is simulated using a VHDL model, to obtain an *observed event trace*. The microarchitectural analysis is modified to predict those events and annotate them to the edges of the generated prediction graph. In this fashion the microarchitectural analysis of an instruction sequence generates a *prediction event graph* that describes an over-approximation of all possible event traces that could occur while executing the instruction sequence. The observed trace of events, the reached execution state, and the consumed time are

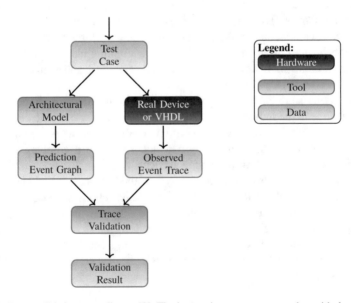

Fig. 2.3 Trace validation according to [9]. The instruction sequences together with the generated prediction graphs annotated by state and timing information are part of the Qualification Support Kit

checked for containment in the prediction graph. Trace validation is successful if the sequence of traced events is found in the prediction graph, and their predicted execution time does not underestimate the observed execution time.

The granularity at which the comparison takes place strongly depends on the debug facilities provided by the hardware. At best, timer interrupts are used to stop execution after each execution cycle. This way, the execution of instruction sequences is extended cycle by cycle to observe actual execution states and execution times.

The behavior of some components of the architectural state, such as the cache state, is unfortunately not directly observable. These need to be indirectly observed through executions that are forced to lead to cache hits and cache misses.

Thus a tremendous effort is required to cover both all instructions and all architectural components. This is essentially achieved by triggering many different architectural states through the execution of dedicated test cases.

The validation suite of the AbsInt static WCET tool aiT may contain several hundred individual test cases, even for a simple DLX-like architecture like the ARM Cortex-M4. For multi-core architectures, such as the TriCore TC275, which features three different cores, several thousand test cases are necessary to cover all architectural features.

How many test cases are required to cover the whole architectural behavior correlates to the complexity of the analyzed hardware, i.e., with the number of available instructions of the instruction set architecture, the number of components of the pipeline architecture like functional units, internal buffers, queues, memories, buses, and their states. Often unexpected (undocumented) hardware behavior is exposed while trying to understand existing test cases. This leads to additional test cases. Hence, the number of test cases that are sufficient in order to cover the (timing) relevant hardware behavior cannot be easily quantified in advance.

2.3 Conclusion

The AbsInt WCET analyzer aiT uses a combination of sound methods to derive safe upper bounds on execution times. Their implementation is quite complex, such that it is natural to query the soundness of the implementation of the technology. We describe the validation efforts employed to convince ourselves, the customers, and the certification authorities of the soundness of the implementation. The European Aviation Safety Agency (EASA), obliged to follow the strictest certification rules, those of DO178-C, has accepted AbsInt's aiT as a validated WCET analysis tool for several time-critical subsystems in the Airbus A380 and A350 planes.

References

1. Rtca/do-178c software considerations in airborne systems and equipment certification (2013)
2. M. Alt, C. Ferdinand, F. Martin, R. Wilhelm, Cache behavior prediction by abstract interpretation. in *Proceedings of the Third International Symposium on Static Analysis, SAS'96*, ed. by R. Cousot, D.A. Schmidt. Aachen, September 24–26, 1996. Lecture Notes in Computer Science, vol. 1145 (Springer, Berlin, 1996), pp. 52–66. https://doi.org/10.1007/3-540-61739-6_33
3. P. Axer, R. Ernst, H. Falk, A. Girault, D. Grund, N. Guan, B. Jonsson, P. Marwedel, J. Reineke, C. Rochange, M. Sebastian, R. von Hanxleden, R. Wilhelm, W. Yi, Building timing predictable embedded systems. ACM Trans. Embedded Comput. Syst. **13**(4), 82:1–82:37 (2014). https://doi.org/10.1145/2560033
4. P. Cousot, R. Cousot, Abstract interpretation: a unified lattice model for static analysis of programs by construction or approximation of fixpoints, in *Conference Record of the Fourth ACM Symposium on Principles of Programming Languages*, ed. by R.M. Graham, M.A. Harrison, R. Sethi, Los Angeles, January 1977 (ACM, New York, 1977), pp. 238–252. https://doi.org/10.1145/512950.512973
5. C. Ferdinand, Cache behaviour prediction for real-time systems. Ph.D. thesis, Saarland University, Saarbrücken (1997)
6. C. Ferdinand, R. Heckmann, M. Langenbach, F. Martin, M. Schmidt, H. Theiling, S. Thesing, R. Wilhelm, Reliable and precise WCET determination for a real-life processor, in *International Workshop on Embedded Software*. Lecture Notes in Computer Science, vol. 2211 (2001), pp. 469–485
7. C. Ferdinand, R. Wilhelm, On predicting data cache behavior for real-time systems, in *Proceedings of the Workshop on Languages, Compilers, and Tools for Embedded Systems (LCTES)*, ed. by F. Mueller, A. Bestavros. Lecture Notes In Computer Science: Languages, Compilers, And Tools For Embedded Systems, vol. 1474 (Springer, Montréal, 1998), pp. 16–30. https://doi.org/10.1007/BFb0057777
8. C. Ferdinand, R. Wilhelm, Efficient and precise cache behavior prediction for real-time systems. Real-Time Syst. **17**(2–3), 131–181 (1999)
9. G. Gebhard, Static timing analysis tool validation in the presence of timing anomalies. Ph.D. thesis, Saarland University (2013). http://scidok.sulb.uni-saarland.de/volltexte/2013/5558/
10. M. Lv, N. Guan, J. Reineke, R. Wilhelm, W. Yi, A survey on static cache analysis for real-time systems. Leibniz Trans. Embedded Syst. **3**(1), 5:1–5:48 (2016). https://doi.org/10.4230/LITES-v003-i001-a005
11. M. Pister, Timing model derivation—pipeline analyzer generation from hardware description languages. Ph.D. thesis, Saarland University (2012)
12. M. Schlickling, Timing model derivation—static analysis of hardware description languages. Ph.D. thesis, Saarland University (2013)
13. H. Theiling, Control flow graphs for real-time systems analysis. Ph.D. thesis, Universität des Saarlandes, Saarbrücken (2002)
14. S. Thesing, Safe and precise WCET determinations by abstract interpretation of pipeline models. Ph.D. thesis, Saarland University (2004)
15. L. Thiele, R. Wilhelm, Design for timing predictability. Real-Time Syst. **28**(2–3), 157–177 (2004). https://doi.org/10.1023/B:TIME.0000045316.66276.6e
16. R. Wilhelm, Mixed feelings about mixed criticality (invited paper), in *Proceedings of the 18th International Workshop on Worst-Case Execution Time Analysis, WCET 2018*, ed. by F. Brandner, July 3, 2018, Barcelona. OASICS, vol. 63 (Schloss Dagstuhl—Leibniz-Zentrum fuer Informatik, 2018), pp. 1:1–1:9. https://doi.org/10.4230/OASIcs.WCET.2018.1

17. R. Wilhelm, S. Altmeyer, C. Burguière, D. Grund, J. Herter, J. Reineke, B. Wachter, S. Wilhelm, Static timing analysis for hard real-time systems, in *Proceedings of the 11th International Conference Verification, Model Checking, and Abstract Interpretation, VMCAI 2010*, ed. by G. Barthe, M.V. Hermenegildo, Madrid, January 17–19, 2010. Lecture Notes in Computer Science, vol. 5944 (Springer, Berlin, 2010), pp. 3–22. https://doi.org/10.1007/978-3-642-11319-2_3
18. R. Wilhelm, D. Grund, J. Reineke, M. Schlickling, M. Pister, C. Ferdinand, Memory hierarchies, pipelines, and buses for future architectures in time-critical embedded systems. IEEE Trans. CAD Integr. Circuits Syst. **28**(7), 966–978 (2009). https://doi.org/10.1109/TCAD.2009.2013287

Chapter 3
The Dynamic Random Access Memory Challenge in Embedded Computing Systems

Matthias Jung, Christian Weis, and Norbert Wehn

3.1 Introduction

Dynamic random access memories (DRAMs) are key components in all computing systems that require large working memory. Due to the strong increase in
data volume in many embedded applications, such as machine learning, image
processing, autonomous systems, etc., DRAMs largely impact the overall system
performance and power consumption. In many of these applications, the overall
system performance is often limited by the memory bandwidth or latency and not by
the computation itself. Due to the dynamic storage scheme of DRAMs and shrinking
technology nodes, reliability is also a major concern in current and future DRAMs.

Therefore, new challenges arise, which we will discuss in this chapter. The
most important metrics, which are typically considered for DRAM subsystems
(especially in the *high-performance computing* (HPC) domain), are *bandwidth*,
latency, and *capacity*. However, in the context of embedded systems it requires
to consider further metrics, such as *power*, *temperature*, *reliability*, *safety*, and
security. In the following we will highlight these challenges and refer to some of
our recent contributions, which tackle these challenges.

M. Jung
Fraunhofer IESE, Kaiserslautern, Germany
e-mail: Matthias.Jung@iese.fraunhofer.de

C. Weis · N. Wehn (✉)
TU Kaiserslautern, Kaiserslautern, Germany
e-mail: weis@eit.uni-kl.de; wehn@eit.uni-klde

© The Author(s) 2021

19

J.-J. Chen (ed.), *A Journey of Embedded and Cyber-Physical Systems*,
https://doi.org/10.1007/978-3-030-47487-4_3

3.2   Bandwidth and Latency

Bandwidth is the amount of data that can be transferred between DRAM and a computational unit within 1 s. The maximum DRAM bandwidth is limited to the number of data pins times the interface frequency. *Latency* is the access time that it takes to complete an access. In fact, latency helps bandwidth, but not vice versa [33]. For instance, lower DRAM latency results in more accesses per second, and therefore higher bandwidth, whereas increasing the number of data pins increases the bandwidth without decreasing latency. A fast execution of applications on embedded systems must not only be supported by the computational units, but the memory subsystem must be designed to avoid hitting the *memory wall* [43]. For example, embedded applications for autonomous driving will require between 400 and 1024 GB/s of memory bandwidth [16], which is hard to realize with the current DRAM technologies. To put the problem in perspective, we survey current memory architectures.

Figure 3.1 shows different DRAM-based memory subsystems, and Figs. 3.2 and 3.3 show their properties with respect to interface frequency, maximum theoretical bandwidth, and energy consumption per transferred bit.[1] The maximum bandwidth of conventional DIMM-based DDR solutions is limited by the I/O count and interface speed. This limitation arises due to the package, power considerations, and costs on both the memory and processing sides.

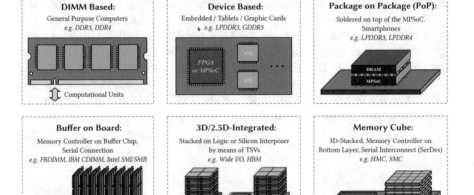

Fig. 3.1 DRAM-based memory subsystems

---

[1]Note that the latency, actual sustainable bandwidth, and the total energy consumption of a DRAM strongly depend on the application being executed. Reaching the maximum theoretical bandwidths in Fig. 3.2 is practically impossible on general-purpose systems.

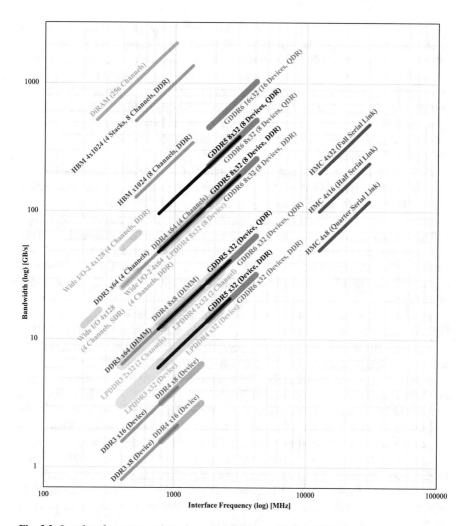

Fig. 3.2 Interface frequency and maximum bandwidth of different DRAM types

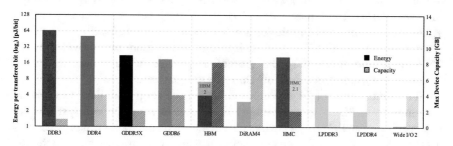

Fig. 3.3 Properties of today's DRAMs (Sources: Micron, Hynix, Nvidia, Xilinx, JEDEC)

To avoid pin limitations, designers and vendors are using *Buffer on Board* (BoB) organizations [7], in which an additional logic component is interposed between the CPU and DRAM to control the memory and to communicate with the CPU over a narrow, high-speed, serial interface. This technique is mainly used in server applications where several terabytes of DRAM are required. The required storage capacity in embedded systems is much smaller than in the high-performance systems BoB targets, and thus this organization is inappropriate. All the other following DRAM devices can achieve easily several GB capacity, which is enough for most of the embedded applications.

Package on Package (PoP) organizations reduce the distance between the DRAM and the MPSoC (from centimeters to millimeters), providing higher bandwidth, lower latency, better power efficiency, and smaller form factors, all of which are especially important for smartphones and tablets. Low power DDR DRAMs (e.g., LPDDR4) can be used either as a device on a PCB or mounted directly as PoP. The latter organization permits only one device to be connected, requiring DRAM commands to be serialized due to the resulting low pin count. For example, if eight LPDDR4 devices are used on a PCB, they deliver a theoretical bandwidth of 137 GB/s.

To address the huge memory demand of highly parallel GPUs, graphic DDR DRAMs (e.g., GDDR5X or GDDR6) use techniques like *quad data rate* (QDR) to deliver high bandwidth compared to conventional DDR DRAM. While LPDDR4 devices are designed and optimized for ultra-low power consumption with aggressive power gating and higher-threshold transistors, GDDR5X/6 devices focus on delivering the highest achievable bandwidth. Both use an architecture with distributed banks (heavy sub-banking) due to the wider data I/O widths of $\times 16/\times 32$ and the larger data prefetch of up to 16 bit per data I/O. However, GDDR5X/6 devices improve the column-to-column cycle time (t_{CCD}) by reducing data path delays from primary sense amplifiers to the global sense amplifiers. Furthermore, GDDR5X/6 chips use an on-die *phase lock loop* (PLL) to achieve very high I/O performance in QDR mode. In contrast, LPDDR4 devices have no on-die PLL or *delay lock loop* (DLL). Combining 16 GDDR6 devices in QDR mode yields a theoretical bandwidth of 1 TB/s, as shown in Fig. 3.2.

Another way of achieving high bandwidth is 3D stacking: examples include WIDE I/O, Micron's *Hybrid Memory Cube* (HMC), and Samsung's *High Bandwidth Memory* (HBM). These memories reduce the distance between CPU and external RAM from centimeters to micrometers by means of *through-silicon via* (TSV) technology. The available bandwidth increases due to more pins provided by the TSVs, but, more importantly, this technology provides a major boost in energy efficiency compared to standard off-chip (G)DDR devices.

The combination of high bandwidth communication and the lower power consumption of 3D integrated memory is an ideal fit for embedded systems. For example, four parallel HBM2 devices on a 2.5D silicon interposer can provide up to 1 TB/s [16]. However, 3D memories suffer from thermal issues, which we discuss in Sect. 3.4.

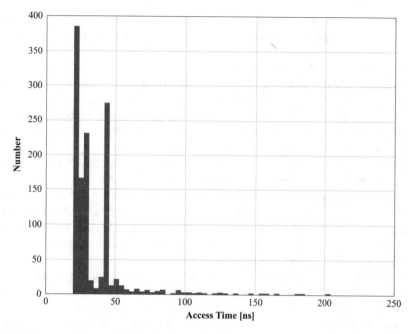

Fig. 3.4 Latency for an application running on DDR3 DRAM

From an application point of view, the DRAM subsystem has non-deterministic timing behavior [8] due to its complex protocol (i.e., the latency of a DRAM request depends on previous issued commands) and the runtime optimization of the memory controller; this makes it difficult to provide predictable performance and thus to guarantee real-time task predictability [1]. Figure 3.4 shows a histogram of the CHStone ADPCM benchmark [11] simulated on the DRAMSys framework [18].[2] Although the average latency is concentrated around 40 ns, the memory latency can easily vary by an order of magnitude.

As with the bandwidth issues discussed above, the memory controller plays an integral role in this non-deterministic timing behavior. The memory controller has to manage, on one side, accesses to the DRAM memory from the compute fabric and, on the other side, the complex interface protocol of the DRAMs. In the following we discuss the main contributions to the DRAM latency that origin from the complex internal memory architecture and the memory controller.

- **Row Misses**: The latency of a bank access varies depending on the state of its row buffer. If a memory access targets the same row as the row currently cached in the buffer (a row hit), it results in lower latency and lower energy memory

---

[2]The simulated DRAM is a DDR3 with a RBC address mapping and disabled scheduler.

accesses. On the other hand, if a memory access targets a different row from that currently in the buffer (a row miss), it results in higher latency and energy consumption.

- **Close vs. Open Page Policy**: *Commercial off-the-shelf* (COTS) DRAM controllers usually support two major modes: an *open page policy* (OPP) and *closed page policy* (CPP). The OPP keeps the current row active after a read or write, whereas the CPP precharges the row automatically after each access. The latter makes the latency for each access more predictable, but it also decreases performance for access patterns with high row-hit potential.
- **Refresh:** DRAMs must be refreshed regularly due to their charge-based bit storage architecture. The memory controller has to issue this refresh operation periodically (e.g., every 64 ms). Normal accesses to the DRAM have to be blocked for the duration of the refresh operation t_{RFC} (350 ns for DDR4), degrading performance with respect to both bandwidth and latency and increasing energy consumption.[3] If a memory access arrives at the same time that a refresh happens it will experience unpredictable latency.
- **Scheduling:** COTS memory controllers are optimized for average case performance and therefore employ runtime scheduling of requests (c.f. Sect. 3.2) for online optimization. For example, with schedulers that attempt to maximize row hits it is possible that a request that misses the row could starve, which again results in a hardly predictable latency.
- **Arbitration:** A major challenge arises when several computational units are issuing read and write requests to the memory controller. The different applications running on these compute units will place their requests in different input buffers, and arbitration must be performed. This leads to interference that can cause high unpredictability.
- **Command/Address and Data bus Contention:** All banks in a DRAM share the same command/address and data buses, which can limit overall performance. If the data bus utilization is 100%, the maximum bandwidth is reached. On the other hand if the command bus utilization is 100%, WR and RD commands must be issued in later cycles that negatively impacts the bandwidth and the latency.
- **Current Limiting and Power Supply Network:** In order to limit peak currents there exists a rolling time-frame, in which a maximum of four banks can be activated, called *four activate window* (t_{FAW}). There is also a minimum time interval between two ACT commands to different banks, (t_{RRD}). Also these constraints can influence bandwidth, latency, and predictability in specific scenarios.
- **Further Effects:** Bank-Groups in DDR4 or GDDR or rank-to-rank switching constraints in DDR memories also impact the predictability.

---

[3]In fact, the degradation grows linearly with the capacity, which means it grows exponentially with each density generation. Liu et al. [27] and Bhati et al. [3] predicted that 40–50% of the power consumption of future DRAM devices will be caused by refresh commands, and the maximum DRAM bandwidth will be significantly reduced.

Due to this unpredictable timing behavior, processors for embedded applications with real-time and strict latency constraints have thus far largely avoided using DRAM. For example, Infineon's Aurix CPU, which is widely used for safety-critical applications, does not provide a DRAM controller.

In past years there were many investigations with respect to DRAM controllers for real-time and mixed-criticality applications in embedded systems. A detailed book which summarizes those approaches has been presented by Goossens et al. [8]. Most of these approaches concentrate operating the DRAM with statically pre-computed command patterns which guarantee a predictable behavior. However, this predictability often comes with a degradation of sustainable bandwidth. Moreover, the bandwidth numbers presented in Fig. 3.2 are theoretical maxima: the sustainable memory bandwidth is much less, and it strongly depends on how the data is stored in the memories, i.e., the memory access pattern [12]. Therefore, it is not only important to choose a memory that provides high bandwidth, it is also important to design a DRAM controller that can bring the sustainable bandwidth closer to the theoretical maximum.

As already mentioned, general-purpose DRAM controllers use online scheduling techniques to improve the sustainable bandwidth, e.g., by reducing the number of row misses or read/write transitions. In order to reduce the number of read/write transitions, DRAM controllers buffer read and write commands in two distinct queues. An arbiter switches between read and write mode to diminish the t_{WTR} penalty, the minimum time interval between the end of a WR burst and a RD command.

However, in embedded systems, many applications (e.g., signal, image, or neural network processing) have regular, fixed, and deterministic memory access patterns. On the compute side, inherent application-specific knowledge has been heavily exploited for efficient compute architectures. However, on the memory side there is limited research that exploits application knowledge to improve the memory access behavior. In [12] we presented an *application-specific memory controller* (ASMC). Key of this controller is an optimized mapping of the logical addresses to physical DRAM addresses such that the row misses in the access pattern stream are minimized. The corresponding mathematical optimization problem is an *integer linear programming* problem. The solution of this problem maximizes the number of row buffer hits and exploits the bank-level parallelism of the DRAM device in order to reduce the latency and therefore to keep up the sustainable bandwidth near to the maximum. Therefore, such an ASMC can outperform online schedulers because it was designed with a *global* application view. Furthermore, for real-time embedded systems with this method we can easily determine WCET bounds, since no non-deterministic online scheduling is involved.

The efficiency of this approach is demonstrated on an industrial embedded image processing application that consists of image rotation and FFT. Due to real-time requirements this application requires a minimum bandwidth of 9.57 GB/s. Figure 3.5 shows the bandwidth and energy for the standard address mappings of a standard memory controller with standard row-bank-column (RBC) mapping and bank-row-column (BRC) mapping, a manual optimization of the mapping of an

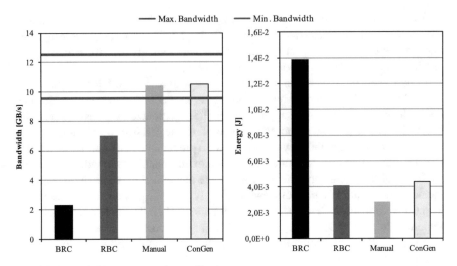

Fig. 3.5 Industrial image processing application

experienced engineer and the ASMC approach. The ASMC approach has a runtime of ~50 min, whereas the manual approach requires ~1 week for an engineer to fully understand the application and analyze the behavior. Furthermore, by using the generated address mapping, all the online scheduling capabilities of the memory controller could be removed, which reduced the required area of the memory controller by 35%.

As mentioned already in Sect. 3.2 the refresh has a large impact on DRAM's bandwidth and latency. The overhead of refreshes can be reduced by only refreshing the memory cells inside the DRAM that hold data that are still alive. A large body of research exists developing schemes that manually refresh the DRAM row-by-row, characterizing each row's ability to retain data and eliminating unnecessary *Refresh* operations on rows that can be refreshed less often. These schemes have been shown to be extremely efficient. Since eliminating refresh improves both energy and performance of the memory system, these schemes offer the potential for significant gains in DRAM-system efficiency. However, these schemes are incompatible with the modern *auto-refresh* mechanism that is widely used: *auto-refresh* operates on multiple rows at once and not on a row-by-row basis. In addition, *auto-refresh* cannot skip any row, whether that row needs to be refreshed or not. Thus, the manual schemes use explicit row-level *Activate* (ACT) and *Precharge* (PRE) commands to refresh row-by-row, called *row granular refresh* (RGR). However, it was shown in [4] that techniques based on RGR could never be as effective as the DRAM's internal auto-refresh.

In [28] we presented a technique called optimized RGR which allows a row-by-row refresh with the same efficiency as the auto-refresh. Here, we investigated the timings that are relevant to *Activate* and *Precharge* commands and showed that these DRAM timing parameters can be reduced for performing the *Refresh*

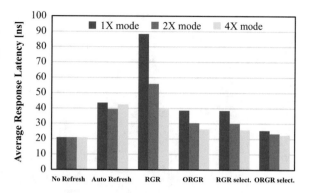

Fig. 3.6 Average response latency using different refresh techniques and modes according to JEDEC: 1X—all rows are refreshed per *Refresh* command, 2X—half of the rows are refreshed, 4X—a quarter of the rows are refreshed

operation row-by-row. We could demonstrate a reduction of latency and increase of bandwidth compared to standard *auto-refresh*, as shown in Fig. 3.6. The results can be even improved if only alive data is refreshed (ORGR select). Additionally, ORGR improved the energy efficiency compared to RGR.

It is becoming clear that embedded applications must concentrate on DRAM solutions like GDDR and HBM in combination with ASMCs and sophisticated refresh mechanisms in order to cope with their high bandwidth and low latency requirements.

3.3   Power Consumption

Power is one of the major challenges in today's embedded system development. According to Fig. 3.3, the preliminary choice for low power designs is LPDDR4 and Wide I/O2 due to their very low energy consumption. However, when aiming at high memory bandwidth, e.g., 1 TB/s these devices are not optimal. For example, to achieve the aforementioned bandwidth with LPDDR4, 64 devices ($\times 32$) are required. Although the average power would be only ~ 17 W at a peak frequency of 2000 MHz, the high number of resulting I/O pins (2048) becomes unfeasible. Hence, the only alternative candidates for high bandwidth are HBM2 and GDDR5X/6. According to Figs. 3.2 and 3.3 the average power consumed[4] by the HBM (4 stack $\times 1024$) and GDDR6 (16 devices, QDR, $\times 32$) devices are ~ 60 W and ~ 150 W, respectively. These numbers show that DRAM will be a significant power contributor to embedded systems which require a high memory bandwidth. Therefore, it is mandatory to efficiently use DRAM's power-down modes in order to reduce power consumption.

In state-of-the-art memory controllers the entry to a power-down mode is scheduled when there was no activity in a period of time called timeout. DRAMs

---

[4]Operated at respective peak frequency.

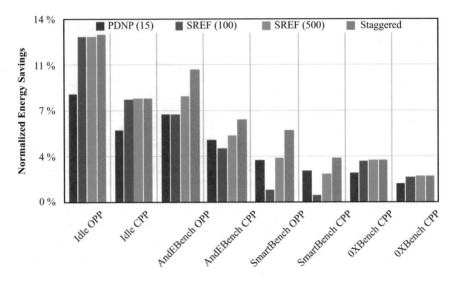

Fig. 3.7 Comparison of energy savings normalized to power-down disabled

offer three power-down modes, called *active power-down* (PDNA), *precharge power-down* (PDNP), and *self-refresh* (SREF). In [35] we showed that a highly opportunistic SREF entry results in an increased power consumption, since the SREF will always execute an internal refresh in the beginning. Therefore, the timeout for a SREF entry should be at least 500 clock cycles for a Wide I/O DRAM.

In [19] we presented an optimized power-down policy, called *staggered power-down*, which considers all three available DRAM power-down modes to achieve the maximum saving in energy and the minimum in slow-down on the execution of the applications. The basic idea is to change to the next more efficient power-down state on a refresh event. With this method, unnecessary SREF entries will be avoided and the hardware timeout counters, as used in state-of-the-art controllers, are not required anymore. As shown in Fig. 3.7 for Wide I/O DRAMs an energy reduction up to 10% in high activity periods and up to 13% in idle phases is feasible.

A high power consumption also fosters a high thermal dissipation that largely impacts the reliability of a DRAM. This challenge is discussed in more detail in the next paragraph.

3.4 Temperature vs. Reliability

DRAMs are very sensitive to high temperature, which increases the leakage in the memory cells. Figure 3.8 shows the different leakage paths in a DRAM cell:

Fig. 3.8 Leakage paths in modern buried wordline DRAM architecture [23, 34]

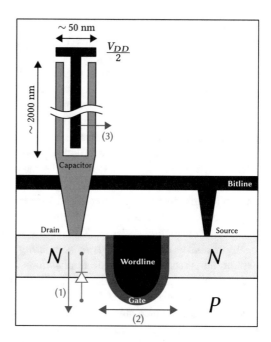

- **Drain Leakage (1)**, which includes the P-N junction leakage as well as *gate induced drain leakage* (GIDL). GIDL is mainly caused by *trap assisted tunneling* (TAT), and it is influenced by the number and distribution of traps in the band-gap region as well as the electric field. Since the negative wordline voltage and the positive charge stored in the cell capacitor (when a 1 is stored in the cell) increase the electric field in the band-gap region (gate-drain overlap region), GIDL is the major source of leakage for a stored 1 in the DRAM cell [31].
- **Sub-threshold Leakage (2)**, which is the drain-source leakage of the cell transistor when it is in the OFF state. This current depends on various factors such as negative wordline voltage, bulk voltage, etc. When the bitlines are in precharged state ($V_{DD}/2$) this can slightly charge the cell capacitor and therefore cause the degradation of a 0 stored in the cell. It can also degrade a 1 stored in the cell by discharging to the bitline, but the leakage will be very small due to the increased threshold voltage of the access transistor when a 1 is stored (body-bias effect).
- **Cell Capacitor Leakage (3)**, which is the leakage through the cell capacitor dielectric. With the technology scaling, also the capacitor area is decreasing. Therefore, to maintain the cell capacitance at the previous value, dielectric thickness has to be reduced, which increases the leakage. The use of new *metal insulator metal* (MIM) structure with high-k dielectric materials has helped to reduce this leakage. Capacitor leakage influences both stored 0's and stored 1's.

In order to avoid data corruption by retention errors due to leakage, the refresh frequency needs to be increased. The general rule of thumb is to double the refresh rate for every 10 °C increase over 85 °C [20]. For example, the refresh period must be decreased from 64 ms to 4–8 ms for 125 °C, which leads to a serious collapse of the sustainable bandwidth [16].

This situation is even worse for today's 3D stacked DRAM systems (e.g., Wide I/O, HBM, HMC, etc.), which aggravate the thermal crisis: i.e., these DRAMs are even more sensitive to temperature changes because of the stacked thin dies. Additionally, when aiming for highest bandwidths with HBM or HMC, these devices will consume, as mentioned before, a significant amount of power on a small area compared to their commodity counterparts. Thus, the self-heating of 3D-DRAMs is even more accelerated. Besides the leakage currents, crosstalk on bitlines and wordlines can also disturb the data stored in the cells or disturb their sensing. Due to the aforementioned effects and shrinking technology nodes, reliability is a major concern in DRAMs. Many techniques exist to improve the reliability, e.g., using *error correcting codes* (ECC) and/or spatial redundancy.

Approximate and *probabilistic computing* evolved as design paradigms that exploit the error resilience of applications to increase their performance and decrease the power consumption [10]. This paradigm can be extended to DRAMs resulting in *approximate DRAMs* (ADRAM) that enable a trade-off between energy efficiency, performance, and reliability. The inherent error resilience of applications allows sacrificing data storage robustness and stability by lowering the refresh rate or disabling refresh in DRAMs completely, as shown in Fig. 3.9. However, to apply ADRAM the statistical DRAM behavior with respect to retention time, process variation, and temperature has to be characterized.

Several studies for the usage of ADRAM are presented in [13–15, 20]. One scenario is safe refresh disabling, i.e., if the data lifetime is smaller than the refresh period, the refresh can be completely switched off without impact on the system

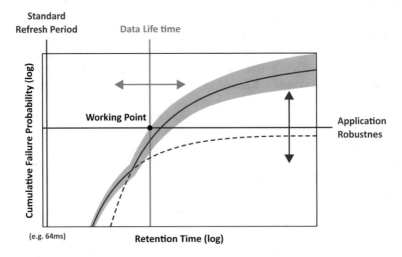

Fig. 3.9 Approximate DRAM design space

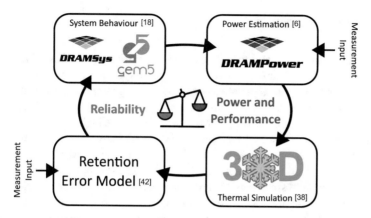

Fig. 3.10 Simulation framework for approximate DRAM explorations

behavior. To perform an accurate characterization we measured state-of-the-art DRAM devices, such as DDR3, DDR4, and Wide I/O. These measurements were the base for a simulation platform for ADRAM investigations.

Exploring ADRAM in a system context is challenging, since a trade-off between accuracy and simulation performance must be considered. Our framework relies on SystemC *Transaction Level Models* (TLM) for fast and accurate simulation. Figure 3.10 shows the closed loop simulation. This simulation loop consists of (1) *DRAM and gem5 Core Models* [5, 18], (2) a DRAM power model [6, 29], which uses either parameters from datasheets, or real measurements [13, 15], (3) a *thermal model* based on 3D-ICE [38], and (4) a *DRAM retention error model* [42].

As mentioned before, ECC is an efficient technique to improve DRAM's reliability, e.g., retention errors or errors induced by crosstalk. State-of-the-art ECC DDR DIMMs, for instance, consist of 8 DRAM devices and a further device for storing the ECC redundancy. Moreover, vendors recently introduced on-die ECC for LPDDR4 [24, 26] to correct retention errors. With ECC the refresh rate can be lowered by 4×, which largely reduces the power consumption. Finding an efficient ECC is a non-trivial task. Traditional ECC techniques for DRAMs assume a symmetric behavior of the retention errors, i.e., the error probability for a stored 0 and 1 is identical. In [22] and [23] we presented a more accurate error model for the retention behavior that exploits the internal cell structure (the so-called true- or anti-cells) of a DRAM. This model is asymmetric and we could show that the channel capacity according to Shannon's capacity definition (the memory cell is considered as a noisy channel) of a single memory cell is larger than in the traditional commonly used symmetrical model. Hence, a more efficient coding must exist. In [23] we presented a new and low-overhead coding scheme that improves the reliability with respect to retention errors.

3.5   Safety and Security

Since DRAMs are more and more used in safety-critical applications like automotive, safety and security are major concerns for DRAMs that were originally mainly developed for consumer applications. Apart from the temperature based retention errors discussed in Sect. 3.4, DRAMs are also prone to transient soft errors, i.e., effects of cosmic particle strikes [9]. Moreover, due to the high frequency of DRAM interfaces transient transmission errors on the DRAM bus can occur. Furthermore, there can be hard errors related to stuck at failures or aging, which could result in a defect column decoder. There exists only a limited amount of studies on DRAM error rates in the field since manufacturers and data centers are very careful to share this sensitive information [30, 36, 41]. Sridharan et al. report 20–66 FIT for a single DRAM device [39, 40]. This highlights the need for appropriate safety mechanisms in order to decrease the FIT rates. For example, the memory controller contains an additional logic that tests the interface periodically in order to detect errors or a strong ECC that is able to correct errors online in order to guarantee functional safety.

Apart from random failures, malicious causes can lead to a safety goal violation, too. Because of transitions to open environments for IoT or *Car2X* communication the vulnerability of DRAMs for embedded systems must be considered. As DRAM process technology scales down, the electrical interference between the memory cells increases, which leads to disturbance errors. Recently, the *row-hammer* problem [21, 32] and its exploits [25, 37] have caused a lot of attention in research and newspapers. By repeatedly opening and closing a DRAM row, called *hammering*, bits in adjacent rows can flip. This effect can be exploited to write on memory locations with prohibited access rights to, e.g., gain kernel privileges or escape a sandbox or hypervisor. In [25] the author showed that secret data can be read with a combination of row-hammer and data dependencies [23]. The row-hammer security attack [21] is a potential malicious behavior that has to be avoided. Controller triggered techniques like *target row refresh* where rows will be refreshed when their activation count exceeds a threshold or techniques on the device level like [2, 44] can alleviate this problem. In [17] a methodology for reverse engineering DRAMs by reconstructing the physical location of memory cells without opening the device package and microscoping the device was presented. This method consists of a retention error analysis while a temperature gradient is applied to the DRAM device. With this insight into the internal DRAM structure row-hammer countermeasure techniques can be improved.

3.6   Conclusion

Emerging applications executed on embedded computing systems require ever increasing main memory sizes. Thus, DRAMs are indispensable to be integrated in such systems. However, the use of DRAMs implies many new challenges.

In this chapter, we highlighted some of the major challenges for the integration of DRAM subsystems into embedded computing systems. These challenges are namely: *bandwidth*, *latency*, *power*, *temperature*, *reliability*, *safety*, and *security*. Furthermore, we showed several solutions from our recent research activities in order to tackle and overcome these challenges.

References

1. A. Agrawal, G. Fohler, DRAM-related challenges in task scheduling with timing predictability on COTS multi-cores for safety-critical Systems, in *Proceedings of the International Symposium on Memory Systems, MEMSYS '17* (ACM, New York, 2017), pp. 265–267. https://doi.org/10.1145/3132402.3132417
2. A. Amaya, H. Gomez, E. Roa, Mitigating Row Hammer attacks based on dummy cells in DRAM, in *2017 IEEE International Conference on Consumer Electronics (ICCE)* (2017), pp. 442–443. https://doi.org/10.1109/ICCE.2017.7889389
3. I. Bhati, M.T. Chang, Z. Chishti, S.L. Lu, B. Jacob, DRAM refresh mechanisms, trade-offs, and penalties. IEEE Trans. Comput. **PP**(99), 1 (2015). https://doi.org/10.1109/TC.2015.2417540
4. I. Bhati, M.T. Chang, Z. Chishti, S.L. Lu, B. Jacob, DRAM refresh mechanisms, penalties, and trade-offs. IEEE Trans. Comput. **65**(1), 108–121 (2016). https://doi.org/10.1109/TC.2015.2417540
5. N. Binkert, B. Beckmann, G. Black, S.K. Reinhardt, A. Saidi, A. Basu, J. Hestness, D.R. Hower, T. Krishna, S. Sardashti, R. Sen, K. Sewell, M. Shoaib, N. Vaish, M.D. Hill, D.A. Wood, The gem5 simulator. SIGARCH Comput. Archit. News **39**(2), 1–7 (2011). https://doi.org/10.1145/2024716.2024718
6. K. Chandrasekar, C. Weis, Y. Li, B. Akesson, O. Naji, M. Jung, N. Wehn, K. Goossens, DRAMPower: open-source DRAM power and energy estimation tool (2014). http://www.drampower.info
7. E. Cooper-Balis, P. Rosenfeld, B. Jacob, Buffer-on-board memory systems, in *2012 39th Annual International Symposium on Computer Architecture (ISCA)* (2012), pp. 392–403. https://doi.org/10.1109/ISCA.2012.6237034
8. S. Goossens, K. Chandrasekar, B. Akesson, K. Goossens, Memory controllers for mixed-time-criticality systems: architectures, methodologies and trade-offs, in *Embedded Systems* (Springer, Berlin, 2016). https://books.google.de/books?id=l9_7CwAAQBAJ
9. M. Greenberg, Understanding automotive DDR DRAM (2017). https://www.synopsys.com/designware-ip/technical-bulletin/automotive-ddr-dram.html
10. J. Han, M. Orshansky, Approximate computing: an emerging paradigm for energy-efficient design, in *2013 18th IEEE European Test Symposium (ETS)* (2013), pp. 1–6. https://doi.org/10.1109/ETS.2013.6569370
11. Y. Hara, H. Tomiyama, S. Honda, H. Takada, Proposal and quantitative analysis of the CHStone benchmark program suite for practical c-based high-level synthesis. J. Inf. Process. **17**, 242–254 (2009). https://doi.org/10.2197/ipsjjip.17.242
12. M. Jung, I. Heinrich, M. Natale, D.M. Mathew, C. Weis, S. Krumke, N. Wehn, ConGen: an application specific dram memory controller generator, in *Proceedings of the Second International Symposium on Memory Systems, MEMSYS '16* (ACM, New York, 2016), pp. 257–267. https://doi.org/10.1145/2989081.2989131
13. M. Jung, D. Mathew, C. Rheinländer, C. Weis, N. Wehn, A platform to analyze DDR3 DRAM's power and retention time. IEEE Design Test **34**(4), 52–59 (2017). https://doi.org/10.1109/MDAT.2017.2705144

14. M. Jung, D. Mathew, C. Weis, N. Wehn, Approximate computing with partially unreliable dynamic random access memory—approximate DRAM, in *Proceedings of the 53rd Annual Design Automation Conference, DAC '16* (ACM, New York, 2016), pp. 100:1–100:4. https://doi.org/10.1145/2897937.2905002

15. M. Jung, D.M. Mathew, C. Weis, N. Wehn, Efficient reliability management in SoCs—an approximate DRAM perspective, in *21st Asia and South Pacific Design Automation Conference (ASP-DAC)* (2016)

16. M. Jung, S.A. McKee, C. Sudarshan, C. Dropmann, C. Weis, N. Wehn, Driving into the memory wall: the role of memory for advanced driver assistance systems and autonomous driving, in *Proceedings of the International Symposium on Memory Systems, MEMSYS '18* (ACM, New York, 2018), pp. 377–386. https://doi.org/10.1145/3240302.3240322

17. M. Jung, C. Rheinländer, C. Weis, N. Wehn, Reverse engineering of DRAMs: Row Hammer with Crosshair, in *International Symposium on Memory Systems (MEMSYS 2016)* (2016)

18. M. Jung, C. Weis, N. Wehn, DRAMSys: a flexible DRAM subsystem design space exploration framework. IPSJ Trans. Syst. LSI Design Methodol. **8**, 63–74 (2015). https://doi.org/10.2197/ipsjtsldm.8.63

19. M. Jung, C. Weis, N. Wehn, M. Sadri, L. Benini, Optimized active and power-down mode refresh control in 3D-DRAMs, in *Proceedings of the 2014 22nd International Conference on Very Large Scale Integration (VLSI-SoC)* (2014), pp. 1–6. https://doi.org/10.1109/VLSI-SoC.2014.7004159

20. M. Jung, E. Zulian, D. Mathew, M. Herrmann, C. Brugger, C. Weis, N. Wehn, Omitting refresh—a case study for commodity and wide I/O DRAMs, in *Proceedings of the 1st International Symposium on Memory Systems (MEMSYS 2015)* (Washington, 2015)

21. Y. Kim, R. Daly, J.H. Kim, C. Fallin, J.H. Lee, D. Lee, C. Wilkerson, K. Lai, O. Mutlu, Flipping bits in memory without accessing them: an experimental study of DRAM disturbance errors, in *ACM/IEEE 41st International Symposium on Computer Architecture (ISCA)* (2014), pp. 361–372. https://doi.org/10.1109/ISCA.2014.6853210

22. K. Kraft, M. Jung, C. Sudarshan, D.M. Mathew, C. Weis, N. Wehn, Improving the error behavior of DRAM by exploiting its Z-channel property, in *IEEE Conference Design, Automation and Test in Europe (DATE)* (2018)

23. K. Kraft, D.M. Mathew, C. Sudarshan, M. Jung, C. Weis, N. Wehn, F. Longnos, Efficient coding scheme for DDR4 memory subsystems, in *ACM International Symposium on Memory Systems (MEMSYS 2018)* (2018)

24. H.J. Kwon, E. Seo, C.Y. Lee, Y.H. Seo, G.H. Han, H.R. Kim, J.H. Lee, M.S. Jang, S.G. Do, S.H. Cho, J.K. Park, S.Y. Doo, J.B. Shin, S.H. Jung, H.J. Kim, I.H. Im, B.R. Cho, J.W. Lee, J.Y. Lee, K.H. Yu, H.K. Kim, C.H. Jeon, H.S. Park, S.S. Kim, S.H. Lee, J.W. Park, S.S. Lee, B.T. Lim, J. Park, Y.S. Park, H.J. Kwon, S.J. Bae, J.H. Choi, K.I. Park, S.J. Jang, G.Y. Jin, 23.4 an extremely low-standby-power 3.733Gb/s/pin 2Gb LPDDR4 SDRAM for wearable devices, in *2017 IEEE International Solid-State Circuits Conference (ISSCC)* (2017), pp. 394–395. https://doi.org/10.1109/ISSCC.2017.7870427

25. A. Kwong, D. Genkin, D. Gruss, Y. Yarom, RAMBleed: reading bits in memory without accessing them, in *Proceedings of the 41st Annual IEEE Symposium on Security and Privacy* (2020)

26. C.K. Lee, Y.J. Eom, J.H. Park, J. Lee, H.R. Kim, K. Kim, Y. Choi, H.J. Chang, J. Kim, J.M. Bang, S. Shin, H. Park, S. Park, Y.R. Choi, H. Lee, K.H. Jeon, J.Y. Lee, H.J. Ahn, K.H. Kim, J.S. Kim, S. Chang, H.R. Hwang, D. Kim, Y.H. Yoon, S.H. Hyun, J.Y. Park, Y.G. Song, Y.S. Park, H.J. Kwon, S.J. Bae, T.Y. Oh, I.D. Song, Y.C. Bae, J.H. Choi, K.I. Park, S.J. Jang, G.Y. Jin, 23.2 a 5Gb/s/pin 8Gb LPDDR4X SDRAM with power-isolated LVSTL and split-die architecture with 2-die ZQ calibration scheme, in *2017 IEEE International Solid-State Circuits Conference (ISSCC)* (2017), pp. 390–391. https://doi.org/10.1109/ISSCC.2017.7870425

27. J. Liu, B. Jaiyen, R. Veras, O. Mutlu, RAIDR: retention-aware intelligent DRAM refresh, in *Proceedings of the 39th Annual International Symposium on Computer Architecture, ISCA '12* (IEEE Computer Society, Washington, 2012), pp. 1–12. http://dl.acm.org/citation.cfm?id=2337159.2337161

28. D.M. Mathew, d.F. Zulian, M. Jung, K. Kraft, C. Weis, B. Jacob, N. Wehn, Using run-time reverse-engineering to optimize DRAM refresh, in *International Symposium on Memory Systems (MEMSYS17)* (2017)

29. D.M. Mathew, E.F. Zulian, S. Kannoth, M. Jung, C. Weis, N. Wehn, A Bank-Wise DRAM power model for system simulations, in *Proceedings of the 9th Workshop on Rapid Simulation and Performance Evaluation: Methods and Tools, RAPIDO '17* (ACM, New York, 2017), pp. 5:1–5:7. https://doi.org/10.1145/3023973.3023978

30. J. Meza, Q. Wu, S. Kumar, O. Mutlu, Revisiting memory errors in large-scale production data centers: analysis and modeling of new trends from the field, in *IEEE/IFIP International Conference on Dependable Systems and Networks (DSN)* (2015)

31. N.J. Min Hee Cho, An innovative indicator to evaluate DRAM cell transistor leakage current distribution. J. Electron Devices Soc. **6**, 494–499 (2017)

32. O. Mutlu, The Row–Hammer problem and other issues we may face as memory becomes denser, in *Design, Automation Test in Europe Conference Exhibition (DATE), 2017*, pp. 1116–1121 (2017). https://doi.org/10.23919/DATE.2017.7927156

33. D.A. Patterson, Latency lags bandwith. Commun. ACM **47**(10), 71–75 (2004). https://doi.org/10.1145/1022594.1022596

34. T. Schloesser, 6F^2 buried wordline DRAM cell for 40 nm and beyond, in *IEEE International Electron Devices Meeting* (San Francisco, 2008)

35. D. Schmidt, N. Wehn, DRAM power management and energy consumption: a critical assessment, in *Proceedings of the 22nd Annual Symposium on Integrated Circuits and System Design* (Natal, 2009)

36. B. Schroeder, E. Pinheiro, W.D. Weber, DRAM errors in the wild: a large-scale field study. ACM SIGMETRICS Perform. Eval. Rev. **37**(1), 193–204 (2009)

37. M. Seaborn, T. Dullien, Exploiting the DRAM Row–Hammer bug to gain kernel privileges (2015). http://googleprojectzero.blogspot.de/2015/03/exploiting-dram-rowhammer-bug-to-gain.html

38. A. Sridhar, A. Vincenzi, M. Ruggiero, T. Brunschwiler, D. Atienza, 3D-ICE: fast compact transient thermal modeling for 3D ICs with inter-tier liquid cooling, in *2010 IEEE/ACM International Conference on Computer-Aided Design (ICCAD)* (2010)

39. V. Sridharan, N. DeBardeleben, S. Blanchard, K.B. Ferreira, J. Stearley, J. Shalf, S. Gurumurthi, Memory errors in modern systems: the good, the bad, and the ugly. SIGARCH Comput. Archit. News **43**(1), 297–310 (2015). https://doi.org/10.1145/2786763.2694348

40. V. Sridharan, D. Liberty, A study of DRAM failures in the field, in *2012 International Conference on High Performance Computing, Networking, Storage and Analysis (SC)* (2012), pp. 1–11. https://doi.org/10.1109/SC.2012.13

41. I. Stefanovici, A. Hwang, B. Schroeder, DRAM's Damning defects—and how they cripple computers. IEEE Spectrum (2015)

42. C. Weis, M. Jung, P. Ehses, C. Santos, P. Vivet, S. Goossens, M. Koedam, N. Wehn, Retention time measurements and modelling of bit error rates of WIDE I/O DRAM in MPSoCs, in *Proceedings of the IEEE Conference on Design, Automation and Test in Europe (DATE)* (European Design and Automation Association, 2015)

43. W.A. Wulf, S.A. McKee, Hitting the memory wall: implications of the obvious. SIGARCH Comput. Archit. News (1995). https://doi.org/10.1145/216585.216588

44. C.M. Yang, C.K. Wei, Y.J. Chang, T.C. Wu, H.P. Chen, C.S. Lai, Suppression of Row Hammer effect by doping profile modification in Saddle-Fin array devices for sub-30-nm DRAM technology. IEEE Trans. Device Mater. Reliab. **16**(4), 685–687 (2016). https://doi.org/10.1109/TDMR.2016.2607174

Chapter 4
On the Formalism and Properties of Timing Analyses in Real-Time Embedded Systems

Jian-Jia Chen, Wen-Hung Huang, Georg von der Brüggen, Kuan-Hsun Chen, and Niklas Ueter

4.1 Introduction

The advanced development of embedded computing devices, accessible networks, and sensor devices has triggered the emergence of complex cyber-physical systems (CPS). In such systems, advanced embedded computing and information processing systems heavily interact with the physical world. Cyber-physical systems are integrations of computation, networking, and physical processes to achieve high stability, performance, reliability, robustness, and efficiency [26]. A cyber-physical system continuously monitors and affects the physical environment which also interactively imposes feedback to the information processing system. The applications of CPS include healthcare, automotive systems, aerospace, power grids, water distribution, disaster recovery, etc.

Due to their intensive interaction with the physical world, in which time naturally progresses, *timeliness* is an essential requirement of correctness for CPS. Communication and computation of safety-critical tasks should be finished within a specified amount of time, called *deadline*. Otherwise, even if the results are correctly delivered from the functional perspective, the reaction of the CPS may be too late and have catastrophic consequences. One example is the release of an airbag in a vehicle, which only functions properly if the bag is filled with the correct amount of air in the correct time interval after a collision, even in the worst-case timing scenario. While in an entertainment gadget a delayed computation result is inconvenient, in the control of a vehicle it can be fatal. Therefore, a modern society cannot adopt a technological advance when it is not safe.

J.-J. Chen (✉) · W.-H. Huang · G. von der Brüggen · K.-H. Chen · N. Ueter
TU Dortmund, Dortmund, Germany
e-mail: Jian-Jia.Chen@tu-dortmund.de; Wen-Hung.Huang@tu-dortmund.de;
Georg.von-der-Brueggen@tu-dortmund.de; Kuan-Hsun.Chen@tu-dortmund.de;
Niklas.Ueter@tu-dortmund.de

© The Author(s) 2021

J.-J. Chen (ed.), *A Journey of Embedded and Cyber-Physical Systems*,
https://doi.org/10.1007/978-3-030-47487-4_4

Cyber-physical systems that require both functional and timing correctness are called cyber-physical real-time systems. Since cyber-physical real-time systems are replacing mechanical and control units that are traditionally operated manually, providing both predictability and efficiency for such systems is crucial to satisfy the safety and cost requirements in our society. Real-time computing for such systems is to provide safe bounds for deterministic or probabilistic timing properties. For providing deterministic timing guarantees, worst-case bounds are pursued. Specifically, the worst-case *execution* time (WCET) of a program (when it is executed exclusively in the system, i.e., without any interference) has to be safely calculated, for details the reader is referred to [31]. The WCETs of multiple programs are then used for analyzing the worst-case *response* time (WCRT) when multi-tasking in the system.

The strongest deterministic timing guarantee ensures that there is no deadline miss of a task by validating whether the WCRT is less than or equal to the specified relative deadline. When the deadlines of all tasks in a system are satisfied, the *hard* real-time requirements are met and the system is a *hard real-time system*. The assumption behind the requirements of hard real-time guarantees is that a deadline miss can result in fatal errors of the system. Ensuring worst-case timing properties has been an important topic for decades. Initially, such worst-case guarantees were achieved by constructing cyclically repetitive static schedules. The timing properties of static schedules can be analyzed easily, but the constructed real-time systems were inflexible to accommodate any upgrades or changes that were not planned in advance.

The seminal work by Liu and Layland [23] provided fundamental knowledge to ensure timeliness and allow flexibility for scheduling periodic real-time tasks in a uniprocessor system. A *periodic task* τ_i is an infinite sequence of task instances, called *jobs*, where two consecutive jobs of a task should arrive recurrently with a period T_i (i.e., the time interval length between the arrival times of two consecutive jobs is always T_i), all jobs of a task have the same *relative* deadline $D_i = T_i$ (i.e., the *absolute* deadline of a job arriving at time t is $t + D_i$), and each job has the same worst-case execution time (WCET) C_i [23]. The *utilization* U_i of a task τ_i is hence defined as $U_i = C_i/T_i$. Although the periodic real-time task model is not always suitable for industrial applications, the exploration of the fundamental knowledge in the past decades provides significant insights. Specifically, Liu and Layland proved the applicability of preemptive dynamic-priority and fixed-priority scheduling algorithms and provided worst-case utilization analysis. To be precise, they showed that as long as the utilization $\sum_{i=1}^{n} C_i/T_i$ of the given n tasks is no more than $n(2^{\frac{1}{n}} - 1)$, which is $\geq 69.3\%$, then the worst-case response time of a task τ_i is guaranteed to be no longer than T_i if the priorities are assigned in the rate-monotonic (RM) order, i.e., τ_i has a higher priority when its period is shorter. Similarly, under preemptive earliest-deadline-first (EDF) scheduling, the utilization bound is guaranteed to be 100%.

However, in many scenarios occasional deadline misses are possible and acceptable. Systems that can still function correctly under these conditions are called *soft* real-time systems. When the deadline misses are bounded and limited, the term *weakly hard* real-time system is used. For such cases, safe and tight quantitative properties of deadline misses have to be analyzed so that the system designers can verify whether the occasional deadline misses are acceptable from the system's perspective. For this purpose, probabilistic timing properties can be very useful, in which the probability of deadline misses or miss rates are pursued. In safety standards, e.g., IEC-61508 and ISO-26262, the probability of failure has to be proved to be sufficiently low. Probabilistic timing properties are important to assure the service level agreements in many applications that require real-time communication and real-time decision-making, such as autonomous driving, smart building, internet of things, and industry 4.0. Deterministic guarantees of interest for weakly hard real-time systems include the quantification of the number of deadline misses within a specified time window length, the worst-case tardiness, and the worst-case number of consecutive deadline misses. Such deadline misses may be allowed and designed on purpose, especially to verify the controller for the physical plant in a CPS. With potential deadline misses in mind, suitable control approaches that can systematically account for data losses can be applied. Such weakly hard real-time systems have been proposed as a feature towards timing-aware control software design for automotive systems in [33].

To design a timing predictable and rigorous cyber-physical real-time system, two separate but co-related problems have to be considered:

1. how to *design scheduling policies* to feasibly schedule the tasks on the platform and system model, referred to as the *scheduler design* problem, and
2. how to *validate* the schedulability of a task system under a scheduling algorithm, referred to as the *schedulability test* problem, to ensure deterministic and/or probabilistic timing guarantees.

The real-time systems research results in the past half-century have a significant impact on the design of cyber-physical systems. Allowing system design flexibility by using dynamic schedules (either fixed-priority or dynamic-priority schedules) has not only academic values but also industrial penetration. Nowadays, most real-time operating systems support fixed-priority schedulers and allow periodic as well as sporadic task activations. When task synchronization or resource sharing is necessary, the priority inheritance protocol and the priority ceiling protocol developed by Sha et al. [27] are part of the POSIX Standards (in POSIX.1-2008).

Existing analyses and optimizations for scheduling algorithms and resource management policies in complex real-time systems are usually *ad-hoc* solutions for a specific studied problem. In this chapter, we challenge this design and analytical practice, since the future design of real-time systems will be more complex, not only in the execution model but also in the parallelization, communication, and synchronization models.

Our Conjecture

We strongly believe that the future design of real-time systems require *formal properties* that can be used modularly to compose safe and tight analysis as well as optimization for the scheduler design and schedulability test problems. This chapter summarizes our recent progress at TU Dortmund for property-based analyses of real-time embedded systems with respect to both deterministic and probabilistic properties.

4.2 Formal Analysis Based on Schedule Functions

For uniprocessor systems, at most one job is executed at a time. Therefore, a **scheduling algorithm** (or **scheduler**) determines the order, in which jobs are executed on the processor, called a schedule. A schedule is an assignment of the given jobs to the processor, such that each job is executed (not necessarily consecutively) until completion. Suppose that $\mathbf{J} = \{J_1, J_2, \ldots J_n\}$ is a set of n given jobs. A schedule for $\mathbf{J}$ can be defined as a function $\sigma : \mathbb{R} \to \mathbf{J} \cup \{\bot\}$, where $\sigma(t) = J_j$ denotes that job J_j is executed at time t, and $\sigma(t) = \bot$ denotes that the system is idle at time t.

If $\sigma(t)$ changes its value at some time t, the processor performs a **context switch** at time t. For a schedule σ to be valid with respect to the arrival time, the absolute deadline, and the execution time of the given jobs, we need to have the following conditions for each J_j in $\mathbf{J}$ for *hard real-time guarantees*:

- $\sigma(t) \neq J_j$ for any $t \leq r_j$ and $t > d_j$ and
- $\int_{r_j}^{d_j} \mathbb{1}_{\sigma(t)=J_j} dt = C_j$, where $\mathbb{1}_{\text{condition}}$ is a binary indicator. If the condition holds, the value is 1; otherwise, the value is 0.

Note that the integration $\int$ of $\mathbb{1}_{\sigma(t)=\text{certain job}}$ over time used in this chapter is only a symbolic representation for summation.

For a given sporadic task set $\mathbf{T}$, each task τ_i in $\mathbf{T}$ can generate an infinite number of jobs as long as the temporal conditions of arrival times of the jobs generated by task τ_i can satisfy the minimum inter-arrival time constraint.

Suppose that the jth job generated by task τ_i is denoted as $J_{i,j}$. Let the set of jobs generated by task τ_i be denoted as $\mathbf{FJ}_i$. A feasible set of jobs generated by a sporadic real-time task τ_i satisfies the following conditions:

- By the definition of the WCET of task τ_i, the actual execution time $C_{i,j}$ of job $J_{i,j}$ is no more than C_i, i.e., $C_{i,j} \leq C_i$.
- By the definition of the relative deadline of task τ_i, we have $d_{i,j} = r_{i,j} + D_i$ for any integer j with $j \geq 1$.
- By the minimum inter-arrival time constraint, we have $r_{i,j} \geq r_{i,j-1} + T_i$ for any integer j with $j \geq 2$.

A feasible set of jobs generated by a periodic real-time task τ_i should satisfy the first two conditions above and the following condition:

- By periodic releases, we have $r_{i,1} = O_i$ and $r_{i,j} = r_{i,j-1} + T_i$ for any integer j with $j \geq 2$.

A *feasible collection* **FJ** *of jobs* generated by a task set **T** is the union of the feasible sets of jobs generated by the sporadic (or periodic) tasks in **T**, i.e., **FJ** $= \cup_{\tau_i \in \mathbf{T}} \mathbf{FJ}_i$. It should be obvious that there are infinite feasible collections of jobs generated by a sporadic real-time task set **T**.

For a feasible collection **FJ** of jobs generated by **T**, a uniprocessor schedule for **FJ** can be defined as a function $\sigma : \mathbb{R} \rightarrow \mathbf{FJ} \cup \{\perp\}$, where $\sigma(t) = J_{i,j}$ denotes that job $J_{i,j}$ is executed at time t, and $\sigma(t) = \perp$ denotes that the system is idle at time t. Recall that we assume that the jobs of task τ_i should be executed in the FCFS manner. Therefore, if $\sigma(t) = J_{i,j}$ then $\sigma(t') \notin \{J_{i,h} | h = 1, 2, \ldots, j - 1\}$, for any $t' > t$ and $j \geq 2$.

The feasibility and optimality of scheduling algorithms should be defined based on all possible feasible collections of jobs generated by **T**.

Definition 4.1 Suppose that we are given a set **T** of sporadic real-time tasks on a uniprocessor system. A schedule σ of a feasible collection **FJ** of jobs generated by **T** is feasible for hard real-time guarantees if the following conditions hold for each $J_{i,j}$ in **FJ**:

- $\sigma(t) \neq J_{i,j}$ for any $t \leq r_{i,j}$ and $t > d_{i,j}$,
- $\int_{r_{i,j}}^{d_{i,j}} \mathbb{1}_{\sigma(t)=J_{i,j}} dt = C_{i,j}$, and
- if $\sigma(t) = J_{i,j}$, then $\sigma(t') \notin \{J_{i,h} | h = 1, 2, \ldots, j - 1\}$, for any $t' > t$ and $j \geq 2$.

A sporadic real-time task set **T** is *schedulable* for hard real-time guarantees under a scheduling algorithm if the resulting schedule of any feasible collection **FJ** of jobs generated by **T** is always feasible. A scheduling algorithm is *optimal* for hard real-time guarantees if it always produces feasible schedule(s) when the task set **T** is schedulable under a scheduling algorithm. □

4.2.1 Preemptive EDF

For the preemptive earliest-deadline-first (EDF) scheduling algorithm, the job in the ready queue whose absolute deadline is the earliest is executed on the processor. To validate the schedulability of preemptive EDF, the *demand bound function* $\mathrm{DBF}_i(t)$, defined by Baruah et al. [1], has been widely used to specify the maximum demand of a sporadic (or periodic) real-time task τ_i to be released and finished in a time interval with length equal to t:

$$\mathrm{DBF}_i(t) = \max\left\{0, \left\lfloor \frac{t - D_i}{T_i} \right\rfloor + 1\right\} \times C_i. \tag{4.1}$$

To prove the correctness of such a demand bound function, we focus on all possible feasible sets of jobs generated by a sporadic/periodic real-time task τ_i. Recall that a feasible set $\mathbf{FJ}_i$ of jobs generated by a sporadic/periodic real-time task τ_i should satisfy the following conditions:

- The actual execution time $C_{i,j}$ of job $J_{i,j}$ satisfies $C_{i,j} \leq C_i$.
- $d_{i,j} = r_{i,j} + D_i$ for any integer j with $j \geq 1$.
- $r_{i,j} \geq r_{i,j-1} + T_i$ for any integer j with $j \geq 2$.

Lemma 4.1 *For a given feasible set $\mathbf{FJ}_i$ of jobs generated by a sporadic/periodic real-time task τ_i, let $\mathbf{FJ}_{i,[r,r+t]}$ be the subset of the jobs in $\mathbf{FJ}_i$ arriving no earlier than r and have absolute deadlines no later than $r + t$. That is,*

$$\mathbf{FJ}_{i,[r,r+t]} = \left\{ J_{i,j} \mid J_{i,j} \in \mathbf{FJ}_i, r_{i,j} \geq r, d_{i,j} \leq r + t \right\}. \tag{4.2}$$

For any r and any $t > 0$,

$$\sum_{J_{i,j} \in \mathbf{FJ}_{i,[r,r+t]}} C_{i,j} \leq \text{DBF}_i(t). \tag{4.3}$$

Proof By definition, $\text{DBF}_i(t) \geq 0$. Therefore, if $\mathbf{FJ}_{i,[r,r+t]}$ is an empty set, we reach the conclusion.

We consider that $\mathbf{FJ}_{i,[r,r+t]}$ is not empty for the rest of the proof. Let $J_{i,j*}$ be the first job generated by task τ_i in $\mathbf{FJ}_{i,[r,r+t]}$. By the definition of $\mathbf{FJ}_{i,[r,r+t]}$ in Eq. (4.2), the arrival time $r_{i,j*}$ of job $J_{i,j*}$ is no less than r, i.e., $r_{i,j*} \geq r$. Since $\mathbf{FJ}_{i,[r,r+t]}$ is not empty, $r_{i,j*} + D_i \leq r + t$.

Since $r_{i,j} \geq r_{i,j-1} + T_i$ for any integer j with $j \geq 2$ for the jobs in $\mathbf{FJ}_i$ as well as the jobs in $\mathbf{FJ}_{i,[r,r+t]}$, the absolute deadlines of the *subsequent* jobs in $\mathbf{FJ}_{i,[r,r+t]}$ are *at least* $r_{i,j*} + T_i + D_i, r_{i,j*} + 2T_i + D_i, r_{i,j*} + 3T_i + D_i, \ldots$. Therefore, there are at most $\left\lfloor \frac{r+t-(r_{i,j*}+D_i)}{T_i} \right\rfloor + 1 \leq \left\lfloor \frac{t-D_i}{T_i} \right\rfloor + 1$ jobs in $\mathbf{FJ}_{i,[r,r+t]}$ since $r \leq r_{i,j*}$. Since the actual execution time $C_{i,j}$ of each job $J_{i,j}$ is no more than C_i by the definition of the jobs in $\mathbf{FJ}_i$, we reach the conclusion. $\qquad\square$

With the help of Lemma 4.1, the following theorem holds.

Theorem 4.1 *A set $\mathbf{T}$ of sporadic tasks is schedulable under uniprocessor preemptive EDF **if and only if***

$$\forall t > 0, \qquad \sum_{\tau_i \in \mathbf{T}} \text{DBF}_i(t) \leq t. \tag{4.4}$$

Proof Only-if part, i.e., the necessary schedulability test. We prove the condition by contrapositive. Suppose that there exists a $t > 0$ such that $\sum_{\tau_i \in \mathbf{T}} \text{DBF}_i(t) > t$, for contrapositive.

For each task τ_i, we create a feasible set of jobs generated by task τ_i by releasing the jobs periodically starting from time 0, and their actual execution times are all set to C_i. By the definition of a uniprocessor system in our scheduling model, at most one job is executed at a time. Therefore, the demand of the jobs that are released no earlier than 0 and must be finished no later than t is strictly more than the amount of available time since $\sum_{\tau_i \in \mathbf{T}} \mathrm{DBF}_i(t) > t$. Therefore, (at least) one of these jobs misses its deadline no matter which uniprocessor scheduling algorithm is used.

Therefore, we can conclude that if the task set $\mathbf{T}$ is schedulable under EDF-P, then $\sum_{\tau_i \in \mathbf{T}} \mathrm{DBF}_i(t) \leq t, \forall t > 0$.

If part, i.e., the sufficient schedulability test: We prove the condition by contrapositive. Suppose that the given task set $\mathbf{T}$ is not schedulable under EDF-P for contrapositive.

Then, there exists a feasible collection of jobs generated by $\mathbf{T}$ which cannot be feasibly scheduled under EDF-P. Let $\mathbf{FJ}$ be such a collection of jobs, where $\mathbf{FJ}_i$ is its subset generated by a sporadic real-time task τ_i in $\mathbf{T}$. Let $\sigma : \mathbb{R} \to \mathbf{FJ} \cup \{\perp\}$ be the schedule of EDF-P for $\mathbf{FJ}$. Since at least one job misses its deadline in σ, let job $J_{k,\ell}$ be the first job which misses its absolute deadline $d_{k,\ell}$ in schedule σ. That is,

$$\int_{r_{k,\ell}}^{d_{k,\ell}} \mathbb{1}_{\sigma(t)=J_{k,\ell}} dt < C_{k,\ell} \leq C_k. \tag{4.5}$$

Let t_0 be the earliest instant prior to $d_{k,\ell}$, i.e., $t_0 < d_{k,\ell}$, such that the processor only executes jobs with absolute deadlines no later than $d_{k,\ell}$ in time interval $(t_0, d_{k,\ell}]$ under EDF-P. That means, immediately prior to time t_0, i.e., $t = t_0 - \epsilon$ for an infinitesimal ϵ, $\sigma(t)$ is either $\perp$ or a job whose absolute deadline is (strictly) greater than $d_{k,\ell}$. We note that t_0 exists since it is at least the earliest arrival time of the jobs in $\mathbf{FJ}$. Moreover, since EDF-P does not let the processor idle unless there is no job in the ready queue, $t_0 \leq r_{k,\ell}$.

Let $\mathbf{FJ}_{i,[t_0,d_{k,\ell}]}$ be the subset of the jobs in $\mathbf{FJ}_i$ arriving no earlier than t_0 and have absolute deadlines no later than $d_{k,\ell}$. That is, we define $\mathbf{FJ}_{i,[t_0,d_{k,\ell}]}$ by setting r to t_0 and t to $d_{k,\ell} - t_0$ in Eq. (4.2). Let $\mathbf{FJ}_{[t_0,d_{k,\ell}]}$ be $\cup_{\tau_i \in \mathbf{T}} \mathbf{FJ}_{i,[t_0,d_{k,\ell}]}$ for notational brevity.

By the definition of t_0, $d_{k,\ell}$, and EDF-P, the processor executes only the jobs in $\mathbf{FJ}_{[t_0,d_{k,\ell}]}$, i.e., $\sigma(t) \in \mathbf{FJ}_{[t_0,d_{k,\ell}]}$ for any $t_0 < t \leq d_{k,\ell}$. Therefore,

$$d_{k,\ell} - t_0 \overset{1}{=} \left(\int_{t_0}^{d_{k,\ell}} \mathbb{1}_{\sigma(t)=J_{k,\ell}} dt \right) + \sum_{J_{i,j} \in \mathbf{FJ}_{[t_0,d_{k,\ell}]} \setminus \{J_{k,\ell}\}} \left(\int_{t_0}^{d_{k,\ell}} \mathbb{1}_{\sigma(t)=J_{i,j}} dt \right)$$

$$\overset{2}{\leq} \left(\int_{t_0}^{d_{k,\ell}} \mathbb{1}_{\sigma(t)=J_{k,\ell}} dt \right) + \left(\sum_{\tau_i \in \mathbf{T}} \sum_{J_{i,j} \in \mathbf{FJ}_{i,[t_0,d_{k,\ell}]}} C_{i,j} \right) - C_{k,\ell}$$

$$\overset{3}{=} \left(\int_{r_{k,\ell}}^{d_{k,\ell}} \mathbb{1}_{\sigma(t)=J_{k,\ell}} dt \right) + \left(\sum_{\tau_i \in \mathbf{T}} \sum_{J_{i,j} \in \mathbf{FJ}_{i,[t_0,d_{k,\ell}]}} C_{i,j} \right) - C_{k,\ell}$$

$$\overset{\text{Eq. (4.5)}}{<} \quad C_{k,\ell} + \left(\sum_{\tau_i \in \mathbf{T}} \sum_{J_{i,j} \in \mathbf{FJ}_{i,[t_0,d_{k,\ell}]}} C_{i,j} \right) - C_{k,\ell}$$

$$\overset{\text{Eq. (4.3)}}{\leq} \quad \sum_{\tau_i \in \mathbf{T}} \mathrm{DBF}_i(d_{k,\ell} - t_0),$$

where the condition $\overset{1}{=}$ is due to $\sigma(t) \in \mathbf{FJ}_{[t_0,d_{k,\ell}]}$ for any $t_0 < t \leq d_{k,\ell}$, the condition $\overset{2}{\leq}$ is due to the definition of a schedule of the jobs in $\mathbf{FJ}_{[t_0,d_{k,\ell}]} \setminus \{J_{k,\ell}\}$, the condition $\overset{3}{=}$ is due to $t_0 \leq r_{k,\ell}$, and $\sigma(t) \neq J_{k,\ell}$ for $t_0 < t \leq r_{k,\ell}$. Hence, there is a certain $\Delta = d_{k,\ell} - t_0$ with $\sum_{\tau_i \in \mathbf{T}} \mathrm{DBF}_i(\Delta) > \Delta$. We reach our conclusion by contrapositive. $\qquad \square$

4.2.2 Preemptive Fixed-Priority Scheduling Algorithms

Under preemptive fixed-priority (FP-P) scheduling, each task is assigned a unique priority before execution and does not change over time. The jobs generated by a task always have the same priority defined by the task. Here, we define $hp(\tau_k)$ as the set of higher-priority tasks than task τ_k and $lp(\tau_k)$ as the set of lower-priority tasks than task τ_k. When task τ_i has a higher priority than task τ_j, we denote their priority relationship as $\tau_i > \tau_j$. We assume that the priority levels are unique.

For FP scheduling algorithms, we need another notation

$$\mathbf{FRJ}_{i,[r,r+\Delta)} = \left\{ J_{i,j} \mid J_{i,j} \in \mathbf{FJ}_i, r_{i,j} \geq r, r_{i,j} < r + \Delta \right\}. \tag{4.6}$$

That is, for a given feasible set $\mathbf{FJ}_i$ of jobs generated by a sporadic/periodic real-time task τ_i, let $\mathbf{FRJ}_{i,[r,r+\Delta)}$ be the subset of the jobs in $\mathbf{FJ}_i$ arriving in time interval $[r, r + \Delta)$. By extending the proofs like in Sect. 4.2.1, we can also prove the following lemma and theorem.

Lemma 4.2 *The total amount of execution time of the jobs of τ_i that are **released** in a time interval $[r, r + \Delta)$ for any $\Delta \geq 0$ is*

$$\sum_{J_{i,j} \in \mathbf{FRJ}_{i,[r,r+\Delta)}} C_{i,j} \leq \left\lceil \frac{\Delta}{T_i} \right\rceil C_i \overset{def}{=} demand_i(\Delta). \tag{4.7}$$

Theorem 4.2 *Let $\Delta_{\min} > 0$ be the minimum value that satisfies*

$$\Delta_{\min} = C_k + \sum_{\tau_i \in hp(\tau_k)} demand_i(\Delta_{\min}). \tag{4.8}$$

The WCRT R_k of task τ_k in a preemptive fixed-priority uniprocessor scheduling algorithm is

- $R_k = \Delta_{\min}$, if $\Delta_{\min} \leq T_k$, and
- $R_k > T_k$, otherwise.

Theorem 4.2 can be re-written into a more popular form, called time-demand analysis (TDA) proposed by Lehoczky et al. [21]: A (constrained-deadline) task τ_k is schedulable under FP-P scheduling if and only if

$$\exists t | 0 < t \leq D_k \leq T_k, \quad C_k + \sum_{\tau_i \in hp(\tau_k)} \left\lceil \frac{t}{T_i} \right\rceil C_i \leq t. \tag{4.9}$$

Theorem 4.2 is a very interesting and remarkable result, widely used in the literature. It suggests to validate the worst-case response time of task τ_k by

- releasing the first jobs of the higher-priority tasks in $hp(\tau_k)$ together with a job of τ_k and
- releasing the subsequent jobs of the higher-priority tasks in $hp(\tau_k)$ as early as possible by respecting their minimum inter-arrival times.

To explain the above phenomena, Liu and Layland in their seminal paper [23] in 1973 defined two terms (according to their wording):

- A **critical instant** for task τ_k is an instant at which a job of task τ_k released at this instant has the largest response time.
- A **critical time zone** for task τ_k is a time interval starting from a critical instant of τ_k to the completion of the job of task τ_k released at the critical instant.

Liu and Layland [23] concluded the famous **critical-instant theorem** as follows: "*A critical instant for any task occurs whenever the task is requested simultaneously with requests for all higher-priority tasks.*" Their proof was in fact incomplete. Moreover, their definition of the critical-instant theorem was incomplete since the condition $\Delta_{\min} > T_k$ was not considered in their definition. A **precise definition of the critical-instant theorem** is revised as follows:

- A **critical instant** for task τ_k is an instant such that

  - a job of task τ_k released at this instant has the largest response time if it is no more than T_k or
  - the worst-case response time of a job of task τ_k released at this instant is more than T_k.

- A **critical time zone** for task τ_k is a time interval starting from a critical instant of τ_k to the completion of the job of task τ_k released at the critical instant.
- In a critical time zone for task τ_k, all the tasks release their first jobs at a critical instant for task τ_k and their subsequent jobs as early as possible by respecting their minimum inter-arrival times.

4.3   Utilization-Based Analyses for Fixed-Priority Scheduling

The TDA in Eq. (4.8) requires pseudo-polynomial-time complexity to check the time points in $(0, D_k]$ for Eq. (4.8), which can be further generalized for verifying the schedulability of task τ_k under fixed-priority scheduling:

$$\exists 0 < t \leq D_k \text{ s.t. } C_k + \sum_{\tau_i \in hp(\tau_k)} \sigma \left(\left\lceil \frac{t}{T_i} \right\rceil C_i + bC_i \right) \leq t, \qquad (4.10)$$

where $\sigma > 0$ and $b \geq 0$. Equation (4.10) can be used in many cases if $D_k \leq T_k$, such as

- $\sigma = 1$ and $b = 0$ in Eq. (4.10) for uniprocessor sporadic task systems [21],
- $\sigma = 1$ and $b = 1$ in Eq. (4.10) for uniprocessor self-suspending sporadic task systems [22] (under the assumption that task τ_k does not suspend itself), and
- $\sigma = 1/M$ and $b = 1$ in Eq. (4.10) for multiprocessor global rate-monotonic scheduling [2] on M identical processors.

Although testing Eq. (4.10) takes pseudo-polynomial time, it is not always necessary to test all possible time points to derive a safe worst-case response time or to provide sufficient schedulability tests. The general and key concept to obtain sufficient schedulability tests in **k2U** in [7, 8] and **k2Q** in [6, 10] is to test only a subset of such points for verifying the schedulability. Traditional fixed-priority schedulability tests often have pseudo-polynomial-time (or even higher) complexity. The idea implemented in the **k2U** and **k2Q** frameworks is to provide a general k-point schedulability test, which only needs to test k points under *any* fixed-priority scheduling when checking schedulability of the task with the kth highest priority in the system. Suppose that there are $k - 1$ higher-priority tasks, indexed as $\tau_1, \tau_2, \ldots, \tau_{k-1}$, than task τ_k. Recall that the task utilization is defined as $U_i = C_i/T_i$. The success of the **k2U** framework is based on a k-point effective schedulability test, defined as follows:

Definition 4.2 (Chen et al. [7, 8]) A k-point effective schedulability test is a sufficient schedulability test of a fixed-priority scheduling policy that verifies the existence of $t_j \in \{t_1, t_2, \ldots t_k\}$ with $0 < t_1 \leq t_2 \leq \cdots \leq t_k$ such that

$$C_k + \sum_{i=1}^{k-1} \alpha_i t_i U_i + \sum_{i=1}^{j-1} \beta_i t_i U_i \leq t_j, \qquad (4.11)$$

where $C_k > 0$, $\alpha_i > 0$, $U_i > 0$, and $\beta_i > 0$ are dependent upon the setting of the task models and task τ_i. □

The properties in Definition 4.2 lead to the following lemmas for the **k2U** framework which are proven in [8].

Lemma 4.3 *For a given k-point effective schedulability test of a scheduling algorithm, defined in Definition 4.2, in which* $0 < t_k$ *and* $0 < \alpha_i \leq \alpha$, *and* $0 < \beta_i \leq \beta$ *for any* $i = 1, 2, \ldots, k - 1$, *task* τ_k *is schedulable by the scheduling algorithm if the following condition holds:*

$$\frac{C_k}{t_k} \leq \frac{\frac{\alpha}{\beta} + 1}{\prod_{j=1}^{k-1}(\beta U_j + 1)} - \frac{\alpha}{\beta}. \tag{4.12}$$

Lemma 4.4 *For a given k-point effective schedulability test of a scheduling algorithm, defined in Definition 4.2, in which* $0 < t_k$ *and* $0 < \alpha_i \leq \alpha$ *and* $0 < \beta_i \leq \beta$ *for any* $i = 1, 2, \ldots, k - 1$, *task* τ_k *is schedulable by the scheduling algorithm if*

$$\frac{C_k}{t_k} + \sum_{i=1}^{k-1} U_i \leq \frac{(k-1)((\alpha + \beta)^{\frac{1}{k}} - 1) + ((\alpha + \beta)^{\frac{1}{k}} - \alpha)}{\beta}. \tag{4.13}$$

Example 4.1 Suppose that $D_k = T_k$ and the tasks are indexed by the periods, i.e., $T_1 \leq \cdots \leq T_k$. When $T_k \leq 2T_1$, task τ_k is schedulable by preemptive rate-monotonic (RM) scheduling if there exists $j \in \{1, 2, \ldots, k\}$ where

$$C_k + \sum_{i=1}^{k-1} C_i + \sum_{i=1}^{j-1} C_i = C_k + \sum_{i=1}^{k-1} T_i U_i + \sum_{i=1}^{j-1} T_i U_i \leq T_j. \tag{4.14}$$

Therefore, the coefficients in Definition 4.2 for this test are $\alpha_i = \beta_i = 1$ and $t_i = T_i$ for $i = 1, 2, \ldots, k - 1$, and $t_k = T_k$. Based on Lemma 4.3, the schedulability of task τ_k under preemptive RM is guaranteed if

$$\frac{C_k}{T_k} \leq \frac{2}{\prod_{j=1}^{k-1}(\beta U_j + 1)} - 1 \quad \Rightarrow \quad \prod_{j=1}^{k}(\beta U_j + 1) \leq 2. \tag{4.15}$$

Based on Lemma 4.4, the schedulability condition of task τ_k under preemptive RM is

$$\sum_{i=1}^{k} U_i \leq k(2^{\frac{1}{k}} - 1). \tag{4.16}$$

The schedulability test in Eq. (4.15) was originally proposed by Bini and Buttazzo [3], called *hyperbolic bound*, as an improvement of the utilization bound in Eq. (4.16) by Liu and Layland in [23]. We note that the original proof in [23] was incomplete, pointed out and fixed by Goossens [15].

The success of the **k2Q** framework is based on a k-point effective schedulability test, defined as follows:

Definition 4.3 A k-point last-release schedulability test under a given ordering π of the $k - 1$ higher-priority tasks is a sufficient schedulability test of a fixed-priority scheduling policy that verifies the existence of $0 \le t_1 \le t_2 \le \cdots \le t_{k-1} \le t_k$ such that

$$C_k + \sum_{i=1}^{k-1} \alpha_i t_i U_i + \sum_{i=1}^{j-1} \beta_i C_i \le t_j, \tag{4.17}$$

where $C_k > 0$, for $i = 1, 2, \ldots, k - 1$, $\alpha_i > 0$, $U_i > 0$, $C_i \ge 0$, and $\beta_i > 0$ are dependent upon the setting of the task models and task τ_i.

The properties in Definition 4.3 lead to the following lemmas for the **k2Q** framework which are proven in [10].

Lemma 4.5 *For a given k-point last-release schedulability test of a scheduling algorithm in Definition 4.3, in which $0 < \alpha_i$, and $0 < \beta_i$ for any $i = 1, 2, \ldots, k-1$, $0 < t_k$, $\sum_{i=1}^{k-1} \alpha_i U_i \le 1$, and $\sum_{i=1}^{k-1} \beta_i C_i \le t_k$, task τ_k is schedulable by the fixed-priority scheduling algorithm if the following condition holds:*

$$\frac{C_k}{t_k} \le 1 - \sum_{i=1}^{k-1} \alpha_i U_i - \frac{\sum_{i=1}^{k-1} (\beta_i C_i - \alpha_i U_i (\sum_{\ell=i}^{k-1} \beta_\ell C_\ell))}{t_k}. \tag{4.18}$$

Example 4.2 Suppose that $D_k = T_k$ and the tasks are indexed by the periods, i.e., $T_1 \le \cdots \le T_k$. When $T_k \le 2T_1$, task τ_k is schedulable by rate-monotonic (RM) scheduling if there exists $j \in \{1, 2, \ldots, k\}$ where

$$C_k + \sum_{i=1}^{k-1} C_i + \sum_{i=1}^{j-1} C_i = C_k + \sum_{i=1}^{k-1} T_i U_i + \sum_{i=1}^{j-1} C_i \le T_j. \tag{4.19}$$

Therefore, the coefficients in Definition 4.3 for this test are $\alpha_i = \beta_i = 1$ and $t_i = T_i$ for $i = 1, 2, \ldots, k - 1$, and $t_k = T_k$. Based on Lemma 4.5, the schedulability of task τ_k under preemptive RM is

$$\frac{C_k}{T_k} \le 1 - \sum_{i=1}^{k-1} U_i - \frac{\sum_{i=1}^{k-1} (C_i - U_i (\sum_{\ell=i}^{k-1} C_\ell))}{T_k}. \tag{4.20}$$

The test in Eq. (4.20) is a quadratic form. The first *quadratic bound* (QB) by Davis and Burns in Equation (26) in [14] and Bini et al. in Equation (11) in [4] is

$$\sum_{i=1}^{k} U_i + \frac{\sum_{i=1}^{k-1} C_i - \sum_{i=1}^{k-1} U_i C_i}{T_k} \le 1. \tag{4.21}$$

The test in Eq. (4.20) is superior to the test in Eq. (4.21).

The generality of the **k2Q** and **k2U** frameworks has been demonstrated in [8, 10]. We believe that these two frameworks, to be used for different cases, have great potential in analyzing many other complex real-time task models, where the existing analysis approaches are insufficient or cumbersome.

For the **k2Q** and **k2U** frameworks, their characteristics and advantages over other approaches have been already discussed in [8, 10]. In general, the **k2U** framework is more precise by using only the utilization values of the higher-priority tasks. If we can formulate the schedulability tests into the **k2U** framework, it is also usually possible to model it into the **k2Q** framework. In such cases, the same pseudo-polynomial-time test is used. When we consider the worst-case quantitative metrics like utilization bounds, resource augmentation bounds, or speedup factors, the result derived from the **k2U** framework is better for such cases. However, there are also cases, in which formulating the test by using the **k2U** framework is not possible. These cases may even start from schedulability tests with exponential-time complexity. We have successfully demonstrated three examples in [6] by using the **k2Q** framework to derive polynomial-time tests. In those demonstrated cases, either the **k2U** framework cannot be applied or with worse results (since different exponential-time or pseudo-polynomial-time schedulability tests are applied).

The automatic procedure to derive the parameters in the **k2U** can be found in [9]. Previously, the parameters in all the examples in [8] were manually constructed. This automation procedure significantly empowers the **k2U** framework to automatically handle a wide range of classes of real-time execution platforms and task models, including uniprocessor scheduling, multiprocessor scheduling, self-suspending task systems, real-time tasks with arrival jitter, services and virtualizations with bounded delays, etc. We believe that the **k2U** framework and the automatic parameter derivations together can be a very powerful tool for researchers to construct utilization-based analyses almost automatically. Depending on the needs of the use scenarios, a more suitable schedulability test class should be chosen for deriving better results.

Utilization-Based Analysis for Dynamic-Priority Scheduling Algorithms
The **k2U** and **k2Q** frameworks provide general utilization-based timing analyses for fixed-priority scheduling. One missing building block is the utilization-based timing analyses for dynamic-priority scheduling algorithms, like EDF. The analytical framework in [8, 10] is based on analytical solutions of linear programming. However, such formulations do not work for EDF.

4.4   Probabilistic Schedulability Tests

In many real-time systems, it is tolerable that at least some of the tasks in the system miss their deadline in rare situations. Regardless, these deadline misses must be quantified to ensure the system's safety. We examine the problem of determining the deadline miss probability of a task under uniprocessor static-priority preemptive scheduling for an uncertain execution behavior, i.e., when each task has distinct execution modes and a related known probability distribution.

One important assumption for real-time systems is that a deadline miss, i.e., a job that does not finish its execution before its deadline, will be disastrous and thus the WCET of each task is always considered during the analysis. Nevertheless, if a job has multiple distinct execution schemes, the WCETs of those schemes may differ significantly. Examples are software-based fault-recovery techniques which rely on (at least partially) re-executing the faulty task instance, mixed-criticality systems, and a reduced CPU frequency to prevent overheating. In all these cases, it is reasonable to assume that schemes with smaller WCET are the common case, while larger WCETs happen rarely.

We use the example of software-based fault-recovery in the following discussion. When such techniques are applied, the probability that a fault occurs and thus has to be corrected is very low, since otherwise hardware-based fault-recovery techniques would be applied. If re-execution may happen multiple times, the resulting execution schemes have an increased related WCET, while the probability decreases drastically. Therefore, solely considering the execution scheme with the largest WCET at design time would lead to largely overdesigning the system resources. Furthermore, many real-time systems can tolerate a small number of deadline misses at runtime as long as these deadline misses do not happen too frequently. This holds true especially if some of the tasks in the system only have weakly hard or soft real-time constraints. Hence, being able to predict the probability of a deadline miss is an important property when designing real-time systems.

We focus on the probability of deadline misses for a single task here, which is defined as follows:

Definition 4.4 (Probability of Deadline Misses) Let $R_{k,j}$ be the response time of the jth job of τ_k. The probability of deadline misses (DMP) of task τ_k, denoted by Φ_k, is an upper bound on the probability that a job of τ_k is not finished before its (relative) deadline D_k, i.e.,

$$\Phi_k = \max_j \left\{ \mathbb{P}(R_{k,j} > D_k) \right\}, \quad j = 1, 2, 3, \ldots. \tag{4.22}$$

It was shown in [24] that the DMP of a job of a constrained- or implicit-deadline task is maximized when τ_k is released at its critical instant. Hence, the time-demand analysis (TDA) in Eq. (4.8) can be applied to determine the worst-case response time of a task when the execution time of each job is known. This implicitly assumes that no previous job has an overrun that interferes with the analyzed job, i.e., we are searching for the probability that the first job of τ_k misses its deadline after a longer interval where all deadlines were met.

When probabilistic WCETs are considered, the WCET obtains a value in $(C_{i,1}, \ldots, C_{i,h})$ with a certain probability $\mathbb{P}_i(j)$ for each job of each task τ_i. Therefore, TDA for a given t is not looking for a binary decision anymore. Instead, we are interested in the probability that the accumulated workload S_t over an interval of length t is at most t. The probability that τ_k cannot finish in this interval is denoted accordingly with $\mathbb{P}(S_t > t)$. The situation where S_t is larger than t is called an *overload* for an interval of length t and hence $\mathbb{P}(S_t > t)$ is the *overload probability* at time t. Since TDA only needs to hold for one t with $0 < t \leq D_k$ to ensure that τ_k is schedulable, the probability that the test fails is upper bounded by the minimum probability among all time points at which the test could fail. As a result, the probability of a deadline miss Φ_k can be upper bounded by

$$\Phi_k = \min_{0 < t \leq D_k} \mathbb{P}(S_t > t). \tag{4.23}$$

The number of points considered in the TDA can be reduced by only considering the *points of interest*, i.e., D_k and the releases of higher-priority tasks.

Therefore, testing the schedulability efficiently requires an efficient routine to calculate $\mathbb{P}(S_t > t)$ for a given t and a combination of given random variables S_t. The research results at TU Dortmund have recently achieved efficient calculations as follows:

- Chernoff bound in [5, 13]: The calculation of $\mathbb{P}(S_t > t)$ is based on the moment generating function of the classical Chernoff bound.
- Multinomial-based approach in [30]: The calculation of $\mathbb{P}(S_t > t)$ uses the multinomial distribution.

We note that the DMP is not identical to the *deadline miss rate* of a task and that the deadline miss rate may be even higher than this probability, as detailed by Chen et al. [11]. However, the approach in [11] utilizes approaches to approximate the deadline miss probability as a subroutine when calculating the rate.

Generality of Using $\mathbb{P}(S_t > t)$
The efficient calculation of $\mathbb{P}(S_t > t)$ results in efficient probabilistic schedulability tests and deadline miss rate analyses for preemptive fixed-priority uniprocessor systems. The general question is whether this holds also for other scheduling problems and platforms, like multiprocessor systems. Whether the applicability can be generalized is an open problem.

4.5   Conclusion

The critical-instant theorem has been widely used in many research results. Some of the extensions of the critical-instant theorem are correct, e.g., the level-i busy window concept in [20], and some are unfortunately incorrect, e.g., for self-suspending tasks in [18, 25]. Specifically, the misconception of modeling self-suspension time of a higher-priority task as its release jitter in the worst-case response time analysis in [25] and [19] had become a standard approach in multiprocessor locking protocols in real-time systems since 2009 until the error was found in 2016, summarized in Section 6 in [12].

In addition to the lack of formalism, the existing properties that have been widely used in analyzing timing satisfactions in cyber-physical real-time systems are also *biased towards computation*. One key assumption used in computation is that the execution of one cycle on a processor reduces the execution of a task by one cycle. If the problem under analysis does not have such a property, the workload characterized by using uniprocessor systems cannot be used at all. To explain this mismatch, consider the preemptive worm-hole switching protocol in communication as an example. Suppose that a message has to be sent from node A to node B by using two switches, called S_1 and S_2. Namely, the message has to follow the path $A \rightarrow S_1 \rightarrow S_2 \rightarrow B$. Suppose that the message is divided into f communication units, in which a communication unit can be sent and received in every time unit. A fast transmission plan is to fully parallelize the communication if possible. That is, one communication unit from A to S_1 for the first time unit, one communication unit from A to S_1 and S_1 to S_2 for the second time units, etc. Therefore, the communication time of the message can be modeled as $f + 2$. This analysis is correct under the assumption that S_1 and S_2 are not used by other flows. However, if the usage of S_1 or S_2 is blocked during the transmission of the message flow, using $f+2$ time units for analysis is problematic. For the fast transmission plan with $f+2$ communication time, it is actually possible that the message is transmitted in $3f$ time units as the links are blocked for any communication parallelism. To handle the increase of time, several factors have been introduced into the real-time analyses for priority-preemptive worm-hole networks, including direct interference, indirect interference, backpressure, non-zero critical instant, sub-route interference, and downstream multiple interference (summarized in Table VII in [17]). However, since the problem under analysis is essentially not the same as a uniprocessor schedule, applying the uniprocessor timing analysis with extensions is in my opinion only possible after a rigorous proof of equivalence. This mismatch leads to a significant amount of flaws in the literature in this topic. Specifically, the analysis in [28] had been considered safe for a few years until a counterexample was provided in [32].

To successfully tackle complex cyber-physical real-time systems that involve computation, parallelization, communication, and synchronization, we believe that new, mathematical, modulable, and fundamental properties for property-based (schedulability) timing analyses and scheduling optimizations are strongly needed.

They should capture the pivotal properties of cyber-physical real-time systems and thus enable mathematical and algorithmic research on the topic. The view angles should not be limited to the processor- or computation-centric perspective. When there are abundant cores/processors, the bottleneck of the system design becomes the synchronization and the communication among the tasks [16, 29]. Different flexibility and tradeoff options to achieve real-time guarantees should be provided in a modularized manner to enable tradeoffs between execution efficiency and timing predictability.

Acknowledgments Part of this work has been supported by European Research Council (ERC) Consolidator Award 2019, PropRT (Number 865170), and Deutsche Forschungsgemeinschaft (DFG), as part of the priority program "Dependable Embedded Systems"—SPP1500, project GetSURE, and the Collaborative Research Center SFB 876 (http://sfb876.tu-dortmund.de/), subprojects A1, A3, and B2.

References

1. S.K. Baruah, A.K. Mok, L.E. Rosier, Preemptively scheduling hard-real-time sporadic tasks on one processor, in *Proceedings of the 11th Real-Time Systems Symposium RTSS*, pp. 182–190 (1990). https://doi.org/10.1109/REAL.1990.128746
2. M. Bertogna, M. Cirinei, G. Lipari, New schedulability tests for real-time task sets scheduled by deadline monotonic on multiprocessors, in *9th International Conference on Principles of Distributed Systems, OPODIS*, pp. 306–321 (2005)
3. E. Bini, G. Buttazzo, G. Buttazzo, A hyperbolic bound for the rate monotonic algorithm, in *13th Euromicro Conference on Real-Time Systems, 2001* (2001), pp. 59–66. https://doi.org/10.1109/EMRTS.2001.bini01
4. E. Bini, T.H.C. Nguyen, P. Richard, S.K. Baruah, A response-time bound in fixed-priority scheduling with arbitrary deadlines. IEEE Trans. Comput. 58(2), 279–286 (2009)
5. K.H. Chen, J.J. Chen, Probabilistic schedulability tests for uniprocessor fixed-priority scheduling under soft errors, in *12th IEEE International Symposium on Industrial Embedded Systems, SIES* (2017), pp. 1–8. https://doi.org/10.1109/SIES.2017.7993392
6. J.J. Chen, W.H. Huang, C. Liu, k2Q: a quadratic-form response time and schedulability analysis framework for utilization-based analysis. CoRR (2015)
7. J.J. Chen, W.H. Huang, C. Liu, k2U: a general framework from k-point effective schedulability analysis to utilization-based tests. CoRR **abs/1501.07084** (2015). http://arxiv.org/abs/1304.1590
8. J.J. Chen, W.H. Huang, C. Liu, k2u: a general framework from k-point effective schedulability analysis to utilization-based tests, in *IEEE Real-Time Systems Symposium, RTSS* (2015), pp. 107–118. https://doi.org/10.1109/RTSS.2015.18
9. J.J. Chen, W.H. Huang, C. Liu, Automatic parameter derivations in *k2U* framework. Computing Research Repository (CoRR) (2016). http://arxiv.org/abs/1605.00119
10. J.J. Chen, W.H. Huang, C. Liu, k2q: a quadratic-form response time and schedulability analysis framework for utilization-based analysis, in *IEEE Real-Time Systems Symposium, RTSS* (2016), pp. 351–362. https://doi.org/10.1109/RTSS.2016.041
11. K.H. Chen, G. von der Brüggen, J.J. Chen, Analysis of deadline miss rates for uniprocessor fixed-priority scheduling, in *24th IEEE International Conference on Embedded and Real-Time Computing Systems and Applications, RTCSA 2018, Hakodate, August 28–31, 2018* (2018), pp. 168–178. https://doi.org/10.1109/RTCSA.2018.00028

12. J.J. Chen, G. Nelissen, W.H. Huang, M. Yang, B. Brandenburg, K. Bletsas, C. Liu, P. Richard, F. Ridouard, N. Audsley, R. Rajkumar, D. de Niz, G. von der Brüggen, Many suspensions, many problems: a review of self-suspending tasks in real-time systems. Real-Time Syst. **55**, 144–207 (2019). https://doi.org/10.1007/s11241-018-9316-9

13. K.H. Chen, N. Ueter, G. von der Bruggen, J.J. Chen, Efficient computation of deadline-miss probability and potential pitfalls, in *Design, Automation & Test in Europe Conference & Exhibition, DATE 2019, Florence, March 25–29, 2019* (2019), pp. 896–901. https://doi.org/10.23919/DATE.2019.8714908

14. R.I. Davis, A. Burns, Response time upper bounds for fixed priority real-time systems, in *Real-Time Systems Symposium, 2008* (2008), pp. 407–418. https://doi.org/10.1109/RTSS.2008.18

15. J. Goossens, Scheduling of hard real-time periodic systems with various kinds of deadline and offset constraints. Ph.D. Thesis, Universite Libre de Bruxelles (1999). http://di.ulb.ac.be/ssd/goossens/Thesis.pdf

16. W.H. Huang, M. Yang, J.J. Chen, Resource-oriented partitioned scheduling in multiprocessor systems: how to partition and how to share? in *Real-Time Systems Symposium (RTSS)* (2016), pp. 111–122

17. L.S. Indrusiak, A. Burns, B. Nikolic, Analysis of buffering effects on hard real-time priority-preemptive wormhole networks. CoRR **abs/1606.02942** (2016). http://arxiv.org/abs/1606.02942

18. K. Lakshmanan, R. Rajkumar, Scheduling self-suspending real-time tasks with rate-monotonic priorities, in *Real-Time and Embedded Technology and Applications Symposium (RTAS)* (2010), pp. 3–12. https://doi.org/10.1109/RTAS.2010.38

19. K. Lakshmanan, D. de Niz, R. Rajkumar, Coordinated task scheduling, allocation and synchronization on multiprocessors, in *Real-Time Systems Symposium (RTSS)* (2009), pp. 469–478. http://dx.doi.org/10.1109/RTSS.2009.51

20. J. Lehoczky, Fixed priority scheduling of periodic task sets with arbitrary deadlines, in *Proceedings Real-Time Systems Symposium (RTSS)* (1990), pp. 201–209. https://doi.org/10.1109/REAL.1990.128748

21. J.P. Lehoczky, L. Sha, Y. Ding, The rate monotonic scheduling algorithm: exact characterization and average case behavior, in *IEEE Real-Time Systems Symposium'89* (1989), pp. 166–171

22. C. Liu, J. Chen, Bursty-interference analysis techniques for analyzing complex real-time task models, in *Real-Time Systems Symposium (RTSS)* (2014), pp. 173–183

23. C.L. Liu, J.W. Layland, Scheduling algorithms for multiprogramming in a hard-real-time environment. J. ACM **20**(1), 46–61 (1973). https://doi.org/10.1145/321738.321743

24. D. Maxim, L. Cucu-Grosjean, Response time analysis for fixed-priority tasks with multiple probabilistic parameters, in *Proceedings of the IEEE 34th Real-Time Systems Symposium, RTSS 2013, Vancouver, December 3–6, 2013* (2013), pp. 224–235. https://doi.org/10.1109/RTSS.2013.30

25. L. Ming, Scheduling of the inter-dependent messages in real-time communication, in *Proceedings of the First International Workshop on Real-Time Computing Systems and Applications* (1994)

26. R. Rajkumar, I. Lee, L. Sha, J. Stankovic, Cyber-physical systems: the next computing revolution, in *Proceedings of the 47th Design Automation Conference* (ACM, New York, 2010), pp. 731–736. https://doi.org/10.1145/1837274.1837461

27. L. Sha, R. Rajkumar, J.P. Lehoczky, Priority inheritance protocols: an approach to real-time synchronization. IEEE Trans. Comput. **39**(9), 1175–1185 (1990). http://dx.doi.org/10.1109/12.57058

28. Z. Shi, A. Burns, Real-time communication analysis for on-chip networks with wormhole switching, in *Proceedings of the Second ACM/IEEE International Symposium on Networks-on-Chip (NOCS)* (2008), pp. 161–170. https://doi.org/10.1109/NOCS.2008.4492735. http://dl.acm.org/citation.cfm?id=1397757.1397996

29. G. von der Brüggen, J.J. Chen, W.H. Huang, M. Yang, Release enforcement in resource-oriented partitioned scheduling for multiprocessor systems, in *Proceedings of the 25th International Conference on Real-Time Networks and Systems, RTNS'17* (ACM, New York, 2017), pp. 287–296. https://doi.org/10.1145/3139258.3139287
30. G. von der Brüggen, N. Piatkowski, K.H. Chen, J.J. Chen, K. Morik, Efficiently approximating the probability of deadline misses in real-time systems, in *Euromicro Conference on Real-Time Systems, ECRTS* (2018), pp. 6:1–6:22. https://doi.org/10.4230/LIPIcs.ECRTS.2018.6
31. R. Wilhelm, J. Engblom, A. Ermedahl, N. Holsti, S. Thesing, D. Whalley, G. Bernat, C. Ferdinand, R. Heckmann, T. Mitra, F. Mueller, I. Puaut, P. Puschner, J. Staschulat, P. Stenström, The worst-case execution-time problem–overview of methods and survey of tools. ACM Trans. Embed. Comput. Syst. **7**(3), 36:1–36:53 (2008). http://doi.acm.org/10.1145/1347375.1347389
32. Q. Xiong, Z. Lu, F. Wu, C. Xie, Real-time analysis for wormhole NoC: revisited and revised, in *2016 International Great Lakes Symposium on VLSI (GLSVLSI)*, (2016), pp. 75–80. https://doi.org/10.1145/2902961.2903023
33. D. Ziegenbein, A. Hamann, Timing-aware control software design for automotive systems, in *Proceedings of the 52Nd Annual Design Automation Conference, DAC'15* (2015), pp. 56:1–56:6. http://doi.acm.org/10.1145/2744769.2747947

Chapter 5
ASSISTECH: An Accidental Journey into Assistive Technology

M. Balakrishnan

5.1 The Beginning: Mainly a Facilitator (2000–2005)

My ASSISTECH journey started by accident. Myself and Mr. Dipendra Manocha (see the adjoining profile) had a common acquaintance with whom he used to meet and exchange audio cassettes in the late 90s and early 2000. Note Dipendra lost his eye sight when he was around 12 years of age. On his reference, Dipendra one day came to meet me with a request that can I help in making the "emacs" editor in Linux accessible through the screen reader software. At that time Dipendra was managing the computer center in NAB (National Association of Blind). At that time I myself was deeply into research and tool development only in the broad areas of VLSI/EDA tools with special focus on system level design that included high level synthesis as well as hardware-software co-design. For such application development, I neither had the skills nor any special interest. On the other hand, I was completely overwhelmed by the sincerity and focus of Dipendra and thus I decided to involve some students through a mini-project. They did manage to develop a basic solution but the solution itself did not get widely deployed but formed the beginning of a very long and fruitful relationship. In the initial years, most of our engagements were similar—he would propose a project (mainly software based) and I would identify a set of project students who would be jointly mentored by us to work on the solution. We became closer and I was always impressed with his ease of understanding the technological capabilities as well as limitations without any formal training in Science or Engineering. Being blind did not seem to matter at all! Note Dipendra's formal college education consists of an under-graduate and post-graduate degree in music. He had immense capability to explain the needs of the visually impaired in a

M. Balakrishnan (✉)
Indian Institute of Technology Delhi, Delhi, India
e-mail: mbala@cse.iitd.ac.in

© The Author(s) 2021
J.-J. Chen (ed.), *A Journey of Embedded and Cyber-Physical Systems*,
https://doi.org/10.1007/978-3-030-47487-4_5

language that could be easily understood by my CSE students. I started inviting him regularly for engaging my CS class to talk about the challenges faced by the VI and these talks were always a big hit and inspired many students.

5.2 Early Phase: Focus on Embedded Systems (2005–2010)

5.2.1 ASSISTECH and COP315

COP315 is a project based embedded systems course developed by me in late 90s. This course has an interesting German connection. I spent a year as a visiting Professor (Konrad Zuse Fellow) in the Computer Science Department at the University of Dortmund in 1994–1995. I came across this semester long course (group project course) where a group of students (upto 10) did a project that typically resulted in a small system development which they demonstrated at the end of the semester. The course had no formal lectures and though they were mentored by the research assistants but primarily were expected to rely on the material available in the open domain for building their backgrounds and solving the problems they encountered. I was impressed by the complexity as well as quality of projects including use of FPGAs, etc. by which they could build their projects and demonstrate. This was in spite of the fact that students at Dortmund had relatively much less exposure through formal instruction in hardware design vis-à-vis IIT Delhi students.

On my return to IIT Delhi, I decided to revamp a course which I used to teach for third year students "Microprocessor based System Design". This course initially had lectures, regular practical assignments as well as a small project. After the change, we made groups of 4–6 students, offered them a list of projects while being open to project suggestions from them as well. As in initial years almost all projects were built around microcontrollers and peripheral devices, I gave them a set of lectures on microcontroller based design. COP315 played a major role in the ASSISTECH journey as it incubated many of the ideas that became successful products later not only in Assistive Technology (AT) space but also in other domains.

5.2.2 SmartCane

The SmartCane journey began as a COP315 project in 2005. This was taken up by a group of 4 students—Rohan Paul, Dheeraj Mehra, Vaibhav Jain, and Ankush Garg who were at that time in the first semester of their third year. The first three students were dual degree students and thus were to stay in the Department for another 3 years. Before the start of the semester, Dipendra first mentioned a key problem in independent mobility of visually impaired in Indian infrastructure—**White cane's inability to detect knee above overhanging obstacles on walking paths without a footprint on the walking path itself. These obstacles can range from low overhanging tree branches on roads and footpaths to jutting out window air-conditioners and room coolers in corridors. This often resulted in upper body injuries and resulted in loss of confidence in independent mobility.**

I am sure that the problem specification meetings the students had with Mr. Manocha inspired them a lot. The unusually intense engagement of this student group led by Mr. Rohan Paul with the objectives of the project was evident from the very beginning. As they worked on the initial prototypes, they also made it known to me that if opportunity is given they would like to work on it even beyond the semester. This group of students continued to work on it for the next 3 years (except Ankush who graduated in 2007), developed a series of prototypes and did multiple user testing at NAB (National Association of Blind) with the help of Mr. Dipendra Manocha. Many of the initial technical challenges were addressed in this phase that lasted from 2005–2008. Except at the very end of this phase, we worked without any external funding but were responsible for building some key collaborations that have lasted more than a decade. First 3-D printing facility has been established in IIT Delhi and was being managed by Prof. P.V.M. Rao in Mechanical Engineering Department. To prototype the casing that was in the form of handle we established contact with Prof Rao. He not only helped them fabricate but got soon involved in various aspects of design. This was the beginning of the relationship that has played a critical role in the success of ASSISTECH. Key technical achievements were also very inter-disciplinary in nature. Apart from innovative use of ultrasonic sensor to get the obstacle distance information and then to convey the same using a vibrator,

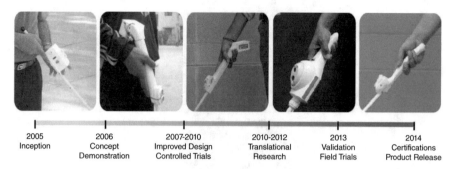

2005	2006	2007-2010	2010-2012	2013	2014
Inception	Concept Demonstration	Improved Design Controlled Trials	Translational Research	Validation Field Trials	Certifications Product Release

Fig. 5.1 SmartCane through various stages of development: Laboratory (2005) to Product (2014)

key technical challenge revolved around reducing power consumption. It was not only in the electronics but also in the coupling between the vibrator and handle for efficient power transfer [2, 18, 19].

At the end of the prototype development (summer of 2007), by a chance coincidence all the three students could get support to present their work at TRANSED 2007 at Montreal. The fact that the paper also got noticed in the conference and received very positive feedback, it helped to motivate the students. Soon after that we were able to sign an agreement with Phoenix Medical Systems (PMS), Chennai for technology transfer and licensing. The period from 2008–2011 was very frustrating as we wrote several proposals for support for translational research to various Government agencies. Challenge was that significant support was also required at the manufacturer's end and agencies were willing to fund only the IIT Delhi Component (Fig. 5.1).

A chance meeting with Dr. Shirshendu Mukherjee, Country Manager of WT (Wellcome Trust, UK) in India resulted in a major success in getting the grant from WT in their first call for "Affordable Health Care in India". Our application itself was successful because the committee was impressed by the team composition— academic entity with product know-how, a credible industrial partner and an NGO in the space as dissemination partner. This clearly was a turning point for the lab. The funding not only catered to the requirements of all the three partners but also included many provisions for creating a quality product, e.g. plastic injection mold costs, large scale multi-city validation trials and costs of getting CE marking. This funding clearly helped us to do many things very systematically and with professional support which is not typical of an academic project. Some of the key features of the development process are listed below.

- A 30-member user group was involved in the early stages of the product design and helped arrive at the requirements
- The product was validated by 150 users in 6 cities before its eventual launch on 31st March 2014. We have not come across any disability product anywhere in the world that has been tested by so many users before its launch

We discuss some of the interesting user-centric design decisions related to ergonomics and aesthetics. Initial designs were difficult for women to grip due to their smaller hand sizes. This came to notice later as the focus group had only men. This required a major redesign including a completely novel packaging. Initially the color of the device was not being considered as it was thought that anyway it is a product for a blind person. A simple question from a blind woman settled this question—do you choose color of the clothes to wear only for yourself or for others as well to see and appreciate?

The success in launching the product within the project duration was highly appreciated by the sponsor (WT) and resulted in significant funding for national and International dissemination. While the industry partner focused on scaling manufacturing, over the next 3 years a team from IIT Delhi and Saksham focused on training users and mobility instructors and building partnership with 40+ agencies across the country. This has played a significant role in widespread acceptance of the device with sales crossing 70,000 units in 5 years (Figs. 5.2 and 5.3).

5.2.3 OnBoard

OnBoard is a globally unique solution for assisting visually impaired users to board public buses independently. It addresses two challenges that are faced by the VI in this process of independent boarding (Figs. 5.4 and 5.5).

Fig. 5.2 SmartCane as a product comes with number of support material for trainers as well as for users (self-learning). It contains manuals in Braille and audio manual in different languages

Fig. 5.3 SmartCane assembly line facility at Phoenix Medical Systems Chennai and a batch of devices packed in a box.

Fig. 5.4 This is a typical bus stop in Delhi. Very often a very large number of routes use the same bus stop and visually impaired person requires the help of another passenger waiting at the stop to help him/her identify the route number

Fig. 5.5 Buses for various reasons do not come and stop in the bay necessitating waiting passengers to walk even up to 25 m to board the bus. For a visually impaired it becomes very difficult as well as unsafe to locate the entry door to board the bus especially as the time available is short and number of passengers may be rushing towards the bus

1. Typically many route numbers use the same bus stop. In Indian Metros in some of the bus stops the number of route numbers being serviced may even go up to 50 but 15+ is quite common and 5+ is almost the norm except in the suburbs. A

VI commuter always requires help of another commuter at the bus stop and that poses two challenges.

- Off-peak time when it is convenient for travel due to buses being less crowded, it may happen that the bus stop has no other passenger waiting
- Even if there are many other passengers waiting, inability of the VI person to choose the right person for information may imply that help is being sought from someone busy (may be on his/her phone) or himself a visitor resulting often in unpleasant situations

2. The second challenge is more typical of bus stop infrastructure and practices followed by buses in picking up the passengers. Buses do not always come and stop in the bay for many reasons including presence of slow moving vehicles or large number of waiting commuters spilling onto the bus lane as well as presence of other buses in the bay. Thus for the VI person to locate the entry door can be even a bigger challenge and our video recordings show that the person may have to walk even up to 25 m.

Technically the work on OnBoard started more or less concurrently with Smart-Cane but after the development of prototypes and some testing, further development was shelved as the focus shifted to translational research on SmartCane. In hindsight it was the right decision as the solution involved changes to the infrastructure provided by a third party (bus operators) struggle has been much higher to get it implemented. The key technical features included a simple user interface, implementation of a slotted network protocol to simultaneously handle up to 8 buses at a bus stop and dual frequency protocol between the user device and the bus device for meeting both the requirements—querying and getting the route number and locating the entry door using an audio cue (Figs. 5.6 and 5.7).

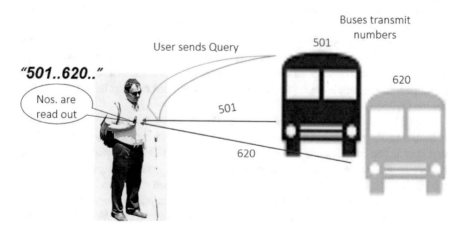

Fig. 5.6 User presses the query button (user module): (1) RF query is sent to all buses in the vicinity and buses respond with their route numbers. (2) User module reads out all the route numbers one by one

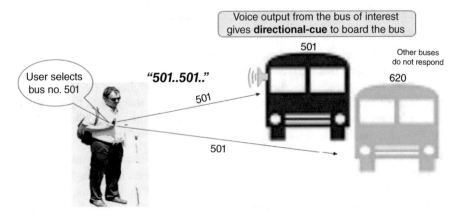

Fig. 5.7 In case a route number is of user interest: (1) User presses the select button (user module), This triggers a voice output from the speaker (bus module fixed near the door). (2) Acts as auditory cue for locating the door

Starting 2014, with the help of TIDE scheme of DST, Govt of India, we restarted this work and did some preliminary testing on DIMTS buses in Delhi. Subsequently with the help of a very active organization in Mumbai (XRCVC[1]) we were able to reach out to BEST where between January and April 2015 we installed these devices on 25 buses (all buses on route numbers 121 and 134 from Back Bay depot). We identified 21 blind bus users, trained them on the device with 5 to 6 supervised boardings each and then asked them to do a total of 350 unsupervised boardings. As they were unsupervised boardings, we needed to evolve a mechanism of assessing the effectiveness of the solution. We asked the users to record the time they reached the bus stop and the time they boarded the bus. Comparing the waiting time at the bus stop with the frequency of the service, we could determine whether the user could board the first bus on the route or not. We achieved 92%+ success rate in users independently boarding the first bus on the route implying the device was very effective. We believe the failure was lower than 8% as sometimes the depot change the buses deployed on a specific route due to operational reasons and it was possible that some of the buses deployed on these routes did not contain the OnBoard bus device [8, 13, 16, 21].

Last 2 years had been spent in reducing the size of the user device and making it aesthetically pleasing. The bus device has also been significantly miniaturized and now it can be operated from the power source of the bus instead of a separate battery as was the case during Mumbai trials Further, a simple protocol by which the device can be retro-fitted in a few minutes has been developed and tested. At present we are awaiting funding for a much larger trials where the users would use the device over a period for their regular commute requirements. We feel such trials

---

[1] http://www.xrcvc.org/.

Fig. 5.8 Mumbai trials on BEST Buses (Jan–April 2015): Bus device was mounted on the window and the unit operated with its own battery. The battery unit is seen below the front seat reserved for the disabled

Fig. 5.9 Miniaturized bus device and user device (not to scale) used in the second Delhi trials. The bus device is mounted on the DMITS orange cluster bus in Delhi. Again the validation trials on these new devices were conducted successfully during May to July 2018

are required before we pursue the same becoming a regulatory requirement so that the bus systems become inclusive which is part of the Government policy (Figs. 5.8 and 5.9).

5.3 Collaborations and Research: Formation of ASSISTECH (2010–2013)

5.3.1 Student Projects to Research

This phase also saw a consolidation of our activities through the formation of ASSISTECH group with a clear objective of working in the space of mobility and education of visually impaired. Allocation of laboratory space in the newly constructed School of Information Technology building significantly facilitated this process. Apart from involvement of students through under-graduate projects, registration of first PhD student (Mr. Piyush Chanana) changed the nature of activities in the laboratory. This also meant that a more structured approach to involving users in evolving product/project specification became possible. Saksham our collaborator posted two of its blind employees to the laboratory and that had an impact at multiple levels. By this time it was clear that any design and product development without involvement of users at all stages, a process that is today referred to as co-creation is essential in this space. It is not sufficient for the designers to understand the limitations of VI users but they also have to understand their strengths in equal measure.

Inter-disciplinary research has gained traction globally but still very often the laboratories draw students from within a single disciplinary background. ASSIS-TECH with its user focus has been able to break this barrier and today has computer science, electrical engineering, mechanical engineering and design expertise apart from user community members under one roof. This has created a unique ecosystem for user-centric approaches to problem solving.

5.3.2 NVDA Activities

NVDA or Non-visual Desk Top access is a screen reader software that makes Microsoft Office products accessible to the visually impaired. It is an open source movement and has been successful in creating a large user base globally. ASSIS-TECH also participated in improving the NVDA tools to essentially support tables that are found in documents. Navigation across the cells of a table in an intuitive and non-verbose manner is critical for VI to comprehend tabular information. ASSISTECH helped augment NVDA in multiple ways and worked closely with the user groups [3, 4].

5.3.3 TacRead and DotBook

Refreshable Braille displays for accessing digital text are a technology which is more than three decades old. These are line display devices that contain an array of cells (typically ranging from 8 to 40) consisting of 6 or 8 pins per cell to create a 6-dot or 8-dot Braille character, respectively. The pins are driven by piezo actuators and through their up and down movement form the Braille characters. Though the devices have been around for a long time but the costs have been so high that there was hardly any penetration in low-income countries like India. Combination of patented technologies, low volume production associated with high margins from monopoly cell manufactures meant the devices continued to cost USD 50 to USD 100 per cell. The rapid growth of screen readers that had started becoming available on all platforms also meant that even in the high-income countries the Braille displays saw a dwindling market. On the other hand, many studies have now shown

 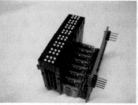 

Fig. 5.10 Initial single Braille cell based on SMA (Shape memory alloys) designed and fabricated during 2014–2017

that if visually impaired persons do not learn to read their ability to write would be minimal. This resulted in the transforming Braille project with worldwide interest in producing low-cost Braille reading and writing devices.[2]

At the same time at ASSISTECH we were engaged in developing refreshable Braille cells using shape memory alloys. The development process involved efforts of multiple batches of mechanical engineering students working with our research staff. These students not only worked on it as their course projects but often stayed back after graduation to work on refining their designs to improve performance or reduce size/weight/cost. It has taken more than 5 years of design, testing, and instrumentation to produce reliable Braille cell modules. Based on the success of the SmartCane, WT funded us again for the Refreshable Braille displays. This time we had two industry partners (PMS and KSPL—KritiKal Solutions Private Limited) and Saksham was again our dissemination partner. The modules produced by Phoenix are being called TacRead, whereas the devices produced by KSPL using these modules are named DotBook. We did a launch of 20-cell and 40-cell devices on Feb 2019. Small volume production is on but some key changes and tooling required for volume production is being setup (Fig. 5.10).

The device is highly complex and represents innovation and design in software, electronics design, mechanical design as well as ergonomics. This also resulted in a number of research papers as well as a well awarded Master's thesis by Suman Muralikrishna [1, 7, 17, 20]. DotBook provides all the key functionality of a laptop including document editing, web browsing as well as emailing. Today all standard

---

[2]http://transformingbraille.org/.

tools have software towards taking inputs from graphical user interfaces. They all needed to be thought afresh as it was not only a character line display but had a length of only 40 characters. Power delivery as well as consumption was a huge challenge as SMA wires required heating to actuate and needed to be cooled to bring the pin return to its original position. Overheating meant not only possibility of wire breakage over time but also slower refresh rate due to higher latency required for deactivation. Mechanical complexity was evident with 320 moving parts (8 pins per cell and 40 cells) which need to be controlled such that the heights of actuated pins are within + or − 0.1 mm in a 1 mm movement. Many user studies were conducted both to identify the functions associated with the limited set of keys that are available as well as their suitable positioning on the device. Volume production is expected to begin by early 2020 as major reliability related issues in the cell module have been sorted out (Fig. 5.11).

5.4   Change of Focus: Technology to Users (2013–2016)

5.4.1   Tactile Graphics Project

During the dissemination phase of SmartCane, it was decided to create a self-learning manual for visually impaired. It was easy to print the Braille text in English and Hindi but the manual also contained a set of simple diagrams to explain ultrasonic ranging. In this process we realized that there is no structured way for producing tactile diagrams in some volume in India. This was in sharp contrast to what we saw in UK and USA where almost all the text material including diagrams were available to visually impaired students as Braille books. Clearly unavailability of tactile diagrams created a significant barrier to visually impaired students to pursue STEM subjects—typically they were forced to study only subjects like history and literature that did not require access to diagrams. This

Fig. 5.11 DotBook: 20-cell and 40-cell devices that were launched on 28 Feb 2019. Volume production is expected to start in first quarter of 2020

prompted us to develop a low-cost technology for production of tactile graphics. Under a project sponsored by MEITY (Ministry of Information Technology), know-how was created for production of low-cost tactile diagrams. This involved some software adaptations, standardization of 3-D printing for preparing low-cost molds, and setting up of facilities for production of tactile material using thermoforming. As tactile route to learning is very different from visual route, guidelines have been developed over many decades in the USA and Europe (e.g. BANA1). A set of tactile designers were trained using these guidelines and then in collaboration with an apex agency in India that is responsible for school curriculum and education, existing Science and Mathematics books for School grades 9th and 10th were produced with tactile diagrams. These were extensively tested with both children studying in blind schools as well as inclusive schools [15]. Once the project was completed, a non-profit company named Raised Lines Foundation[3] (RLF) has been incubated in August 2018. RLF is at present engaged in design and production of tactile diagrams and other tactile material for education. Initial feedback suggests that it is helping many visually impaired students to pursue STEM subjects all over India. In the next few years not only we intend to scale this venture but also create partnerships across the country for creation of tactile material in different Indian languages. It is also planned to create a design service for organizations outside India (Figs. 5.12 and 5.13).

5.4.2   More Research Projects and International Collaboration

This period also saw enrollment of number of PhD and Masters students in this space. As Design students started enrolling with us, it was clear that there were many

Fig. 5.12 Tactile diagrams developed as part of the project Tactile Graphics sponsored by MEITY, Govt of India made innovative use of 3-D printing for production of low-cost tactile diagrams

---

[3]http://raisedlines.org/.

Fig. 5.13 A non-profit company has been formed using the know-how developed under the project to produce school text books and other reading material for visually impaired. The company is in operation since August 2018

open questions to be answered to make the tactile diagrams effective. Also modern tactile production processes have created more flexibility in production but there have been little study on their effectiveness. Ms. Richa Gupta, a design graduate, started studying effectiveness of various forms of representation of tactile diagrams and is now close to finishing her work [10, 14].

This research on tactile diagrams also created our first significant collaboration with IUPUI in Indianapolis. Prof. Steven Mannheimer, whom we met in a Indo-US workshop in India was also interested in pursuing this area. He had already done some work with Indianapolis School for the Blind (ISB). This collaboration enabled Richa to conduct her studies and establish effectiveness of representation techniques in two distinct geographies—NAB in New Delhi and ISB at Indianapolis [11].

5.5   Consolidation and Growth (2016 -)

The period after 2016 has seen lot of growth—both in terms of research projects, students as well as activities. This period also saw lots of national and International recognitions.

5.5.1   RAVI

Reading Assistant for Visually Impaired (RAVI) is a project aimed at making pdf documents accessible. The work is under progress and we expect some initial results next year [6, 12]. The challenge is manifold

- The legacy documents available in the digital library in Indian languages use fonts that are not recognized by screen reader software. There is a need to convert these into formats like ePUB that are accessible.
- Mathematics still poses a major challenge as very often in pdf documents the equations are available as images. Even otherwise delivering a complex equation in audio format that is linear and comprehensible is a challenge. One of the visually impaired students in the group (Mr. Akashdeep Bansal) has taken it up as his PhD research topic. We are also collaborating with Prof. Volker Sorge in University of Birmingham (UK) who has had extensive experience in this field.
- Navigating through tables efficiently needs some research as well as tooling.
- Diagrams require associated description for delivery. Recent AI techniques are making great progress in automatically describing images and we would like to adapt these techniques for automatic generation of diagram descriptions

5.5.2 MAVI

Mobility Assistant for Visually Impaired (MAVI) is a project to use modern AI based image classification techniques for safe and efficient mobility of visually impaired. The focus of this work is to look at object detection in the context of street infrastructure that is typical of Indian cities. The initial prototypes have been built but a usable solution is likely only by 2021. The video stream from the camera is processed using multiple streams to detect

- Street stray animals like dogs and cattle for safety
- Potholes at a distance again for safety
- Multi-lingual street signage for assisting in navigation
- Face detection for social inclusion

5.5.3 NAVI

Navigation Assistant for Visually Impaired (NAVI) is a mobile app based solution that can help in outdoor as well as indoor mobility. Mr. Piyush Chanana, a senior scientist in ASSISTECH, understood the challenges of VI persons in independent mobility by interacting with hundreds of blind users whom he has trained in the use of SmartCane. He has captured this in the specification and design of an app that can help visually impaired in outdoor mobility. Among other things, it involves annotation of tactile landmarks that are "visible" to a blind user [5, 9].

Currently another PhD student, Mr. Vikas Upadhyay has started working on indoor navigation. The work involves effective mapping of internal spaces, localization with limited additional infrastructure, and an appropriate user interface for visually impaired. Intent is not only to propose novel algorithms and techniques but also to install it in couple of public buildings and validate.

5.5.4 Outreach Through Conferences

This period also saw our intent to outreach and create forums for all Assistive Technology stakeholders to come together. Initially we organized two Indo-US workshops with participation of 50+ persons in this space. This was followed by two major assistive technology conferences in 2018 and 2019. EMPOWER 2018[4] and EMPOWER 2019[5] brought 250+ AT researchers, users, innovators and entrepreneurs, educators, exhibitors, etc. on one platform and have been a major success.

---

[4]http://assistech.iitd.ac.in/empower2018/.

[5]http://assistech.iitd.ac.in/empower2019/.

5.5.5 Major Recognitions

This period also saw numerous awards being conferred for ASSISTECH activities. A spotlight talk in London as part of the Grand Challenges[6] meeting organized by Gates foundation and Wellcome Trust in London on 24th Oct 2016 was the first major recognition of ASSISTECH activities. Major Indian awards included NCPEDP-Mphasis Universal Design award and three national awards. ACM recognized our contribution through the ACM Eugene L Lawler[7] award for Humanitarian Contributions within Computer Science and Informatics at their annual awards function in San Francisco on 15th June 2019.

5.6 Conclusion

Clearly my venturing into Assistive Technology from Embedded Systems/EDA space has been an accident and thus I titled this paper as an accidental journey. The work has been hugely satisfying primarily because of the impact it potentially has on the lives of visually impaired people. The feedback we frequently get from our numerous users on how our devices and solutions have positively affected their lives is sufficient to drive and inspire us. Over a period ASSISTECH design philosophy has become completely user-centric. In the initial years we used to choose the problems to look at based on our own experience and expertise. Now if we learn of a major challenge in the visually impaired community and then scout around and try to put up a team by collaborating with people with the required expertise. Our motto is to touch a million people by 2022.

References

1. A Compliant Mechanism Design for Refreshable Braille Display Using Shape Memory Alloy, *International Design Engineering Technical Conferences and Computers and Information in Engineering Conference*, vol. 9. ASME/IEEE International Conference on Mechatronic and Embedded Systems and Applications (2015)
2. M. Balakrishnan, K. Paul, A. Garg, R. Paul, D. Mehra, V. Singh, P. Rao, V. Goel, D. Chatterjee, D. Manocha, Cane mounted knee-above obstacle detection and warning system for the visually impaired, in *3rd ASME/IEEE International Conference on Mechatronic and Embedded Systems and Applications (MESA 2007)* (2007)
3. M. Belani, S. Gupta, D. Kaushal, M. Balakrishnan, Microsoft excel chart accessibility: an affordable and effective solution, in *Digitization and E-Inclusion in Mathematics and Science 2016 (DEIMS 2016)* (2016)

---

[6]https://grandchallenges.org/video/smartcane-m-balakrishnan-india.
[7]https://awards.acm.org/lawler.

4. M. Belani, D. Kaushal, M. Agrawal, M. Balakrishnan, Describeit, in *CSUN Assistive Technology Conference* (2018)
5. P. Chanana, R. Paul, M. Balakrishnan, P. Rao, Assistive technology solutions for aiding travel of pedestrians with visual impairment. J. Rehab. Assist. Technol. Eng. **4**, 2055668317725, 993 (2017)
6. S.P. Chowdhary, D. Manocha, M. Balakrishnan, A. Bansal, H. Garg, Making legacy digital content accessible at source, in *Proceedings of the 16th Web For All 2019 Personalization – Personalizing the Web, W4A '19* (ACM, New York, 2019), pp. 13:1–13:2. https://doi.org/10.1145/3315002.3332444
7. Design and Comparative Analysis of Linear Guides for Refreshable Braille Displays, *International Design Engineering Technical Conferences and Computers and Information in Engineering Conference*, vol. 5B. 41st Mechanisms and Robotics Conference (2017)
8. M. Dheeraj, M. Balakrishnan, et al., Design for user testing of affordable bus identification and homing system for the visually impaired, in *TRANSED 2012* (013)
9. J.D. Dhruv, A. Prabhav, V. Maheshwari, M. Aman, Gupta, Ollagnier, Goyal, M.A. Paul, Ashish, Paldhe, Daş, Sanyal, Taneja, Manocha, M.S. Rao, Design and user testing of an affordable cellphone based indoor navigation system for visually impaired, in *TRANSED 2012* (2012)
10. R. Gupta, M. Balakrishnan, P.V.M. Rao, Tactile diagrams for the visually impaired. IEEE Potentials **36**(1), 14–18 (2017). https://doi.org/10.1109/MPOT.2016.2614754
11. R. Gupta, P. Rao, M. Balakrishnan, S. Mannheimer, Basic identity tags (bits) in tactile perception of 2d shape. J. Technol. Person Disabil. **6**, 103 (2018)
12. S. Holani, A. Bansal, M. Balakrishnan, Pushpak: Voice command-based ebook navigator, in *Proceedings of the 16th Web For All 2019 Personalization – Personalizing the Web, W4A '19* (ACM, New York, 2019), pp. 14:1–14:2. https://doi.org/10.1145/3315002.3332445
13. D. Jain, A. Jain, R. Paul, A. Komarika, M. Balakrishnan, A path-guided audio based indoor navigation system for persons with visual impairment, in *Proceedings of the 15th International ACM SIGACCESS Conference on Computers and Accessibility, ASSETS '13* (ACM, New York, 2013), pp. 33:1–33:2. https://doi.org/10.1145/2513383.2513410
14. K. Kunal, K. Renu, V. Lipika, C. Vibha, A. Mayank, C. Kameshwar, M. Balakrishnan, Converting mathematics textbook to tactile form: process and experiences, in *DEIMS 2016* (2016)
15. M. Mech, K. Kwatra, S. Das, P. Chanana, R. Paul, M. Balakrishnan, Edutactile – a tool for rapid generation of accurate guideline-compliant tactile graphics for science and mathematics, in *Computers Helping People with Special Needs*, ed. by K. Miesenberger, D. Fels, D. Archambault, P. Peňáz, W. Zagler (Springer International Publishing, Cham, 2014), pp. 34–41
16. D. Mehra, D. Gupta, T. Vishwarath, N. Shah, P. Chanana, Siddharth, R. Paul, M. Balakrishnan, P. Rao, Bus identification system for visually impaired: Evaluation and learning from field trials, in *TRANSED 2015* (2015)
17. A.S. Muralikrishnan, P. Sapra, S. Agrawal, P. Chanana, M. Balakrishnan, P.V.M. Rao, FPGA-based controllers for compact low power refreshable braille display, in *2018 IEEE Computer Society Annual Symposium on VLSI (ISVLSI)* (2018), pp. 632–637. https://doi.org/10.1109/ISVLSI.2018.00120
18. R. Paul, A. Garg, V. Singh, D. Mehra, M. Balakrishnan, K. Paul, D. Manocha, Smart cane for the visually impaired: technological solutions for detecting know above obstacles and accessing public buses, in *Proceedings of 11th International Conference on Mobility and Transport for Elderly and Disabled Persons (TRANSED 2007)* (2007)
19. R. Paul, et al., M.B.: smart cane for the visually impaired, in *TRANSED 2010* (2010)
20. Refreshable Braille Display Using Shape Memory Alloy With Latch Mechanism, *International Design Engineering Technical Conferences and Computers and Information in Engineering Conference*, vol. 9. 13th ASME/IEEE International Conference on Mechatronic and Embedded Systems and Applications (2017)
21. V. Sharma, et al., M.B.: user triggered bus identification and homing system: making public transport accessible for the visually challenged, in *TRANSED 2010* (2010)

Chapter 6
Reflecting on Self-Aware Systems-on-Chip

Bryan Donyanavard, Tiago Mück, Kasra Moazzemi, Biswadip Maity,
Caio Batista de Melo, Kenneth Stewart, Saehanseul Yi, Amir M. Rahmani,
and Nikil Dutt

6.1 Introduction to Self-Aware Systems-on-Chip

We are seeing an increasing number of complex cyber-physical systems (CPS) deployed for various applications, such as road-traffic control involving communicating autonomous cars and infrastructure, or smart grids controlling energy delivery down to the individual device. These distributed applications follow common design objectives, such as energy-efficiency, and require guarantees for high availability, real time or safety. In this context, autonomy is crucial: multiple system goals varying over time need to be adaptively managed and objectives holistically coordinated. By empowering future CPS with self-awareness, these systems promise to dynamically adapt, learn, and manage unforeseen changes [6].

6.1.1 Computational Self-Awareness

Computational self-awareness is the ability of a computing system to recognize its own state, possible actions, and the result of these actions on itself, its operational goals, and its environment, thereby empowering the system to become autonomous [6]. Computational self-awareness in itself is not a new field, but

---

B. Donyanavard (✉) · T. Mück · K. Moazzemi · B. Maity · C. B. de Melo · K. Stewart · S. Yi · N. Dutt
University of California, Irvine, CA, USA
e-mail: bdonyana@uci.edu; tmuck@uci.edu; moazzemi@uci.edu; maityb@uci.edu; cbatista@uci.edu; kennetms@uci.edu; saehansy@uci.edu; dutt@uci.edu

A. M. Rahmani
Technische Universität Wien, Vienna, Austria
e-mail: amirr1@uci.edu

© The Author(s) 2021
J.-J. Chen (ed.), *A Journey of Embedded and Cyber-Physical Systems*,
https://doi.org/10.1007/978-3-030-47487-4_6

rather a unification of subjects studied disjointly in various fields including control systems, artificial intelligence, autonomous computing, software engineering, among others, and how such research can be applied toward building computer systems with varying degrees of self-awareness in order to accomplish a task [8].

6.1.2   Cyber-Physical Systems-on-Chip

Battery-powered devices are the most ubiquitous computers in the world. Users of battery-powered devices expect support for various high-performance applications running on same device, potentially at the same time. Applications range from interactive maps and navigation, to web browsers and email clients. In order to meet performance demands by users utilizing complex workloads, increasingly powerful hardware platforms are being deployed in battery-powered devices. Systems-on-chip (SoCs) can integrate hundreds of heterogeneous cores and uncore components on a single chip. Such systems are constrained by a limited amount of shared system resources (e.g., power, interconnects). Simultaneously, the systems are expected to support workloads with diverse characteristics and demands that may conflict with system constraints. These platforms include a number of configurable knobs throughout the system stack and with different scope that allow for a trade-off between power and performance, e.g., dynamic voltage and frequency scaling (DVFS), power gating, idle cycle injection. These knobs can be set and modified at runtime based on the workload demands and system constraints. Heterogeneous many-core processors (HMPs) have extended this principle of dynamic power-performance trade-offs by incorporating single-ISA, architecturally differentiated cores on a single processor, with each of the cores containing a number of independent trade-off knobs. All of these configurable knobs allow for a huge range of potential trade-off. With such a large number of possible configurations, SoCs require intelligent runtime management in order to achieve system goals for complex workloads. Additionally, the knobs may be interdependent, so the decisions must be coordinated.

Cyber-physical systems-on-chip (CPSoC) [21] provide an infrastructure for system introspection and reflective behavior, which is the foundation for computational self-awareness. Figure 6.1 shows the infrastructure of a sensor-actuator rich platform, integrated with decision-making entities that observe system state through virtual and physical sensors at various layers in order to set the system configuration through actuators. The actuations are determined by policies that enforce the overall application goals while considering system constraints. Such an infrastructure can deploy *re*active policies through the traditional *Observe, Decide,* and *Act* (ODA) feedback loop, as well as *pro*active policies through the augmented *self-aware* feedback loop. Figure 6.2 shows how the traditional ODA loop is augmented with reflection to provide self-aware adaptation. In this chapter we explore the

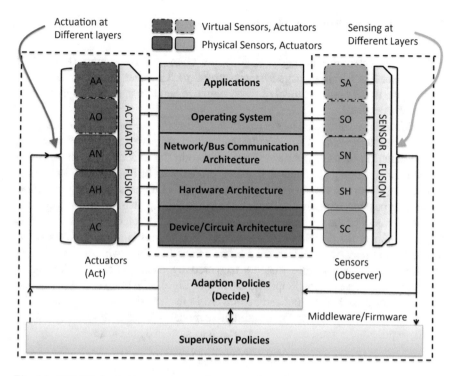

Fig. 6.1 CPSoC infrastructure: sensors and actuators throughout the system stack, with support for adaptive policies that enforce a given goal (from [3])

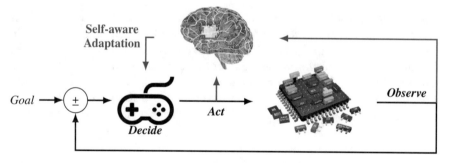

Fig. 6.2 Self-aware feedback loop. Policies are deployed to make action decisions toward achieving a goal by controlling the CPSoC based on observations and self-aware adaptation

use of computational self-awareness to address challenges of adaptive resource management in cyber-physical systems-on-chip.[1]

---

[1]Throughout the remainder of this chapter we use SoC as an umbrella term that includes CPSoC.

6.2   Reflective System Models

Traditionally, resource managers deploy an ODA feedback loop (lower half (in black) of Fig. 6.3) to manage systems at runtime. However, recent works [1, 27] have shown that a runtime model of the system can better manage the unpredictable nature of workloads.

Reflection can be defined as *the capability of a system to reason about itself and act upon this information* [26]. A reflective system can achieve this by maintaining a representation of itself (i.e., a self-model) within the underlying system, which is used for reasoning. Reflection is a key property of self-awareness. Reflection enables decisions to be made based on both *past* observations, as well as *predictions* made from past observations. Reflection and prediction involve two types of models: (1) a self-model of the subsystem(s) under control, and (2) models of other policies that may impact the decision-making process. Predictions consider *future* actions, or events that may occur before the next decision, enabling "what-if" exploration of alternatives. Such actions may be triggered by other policies invoked more frequently than the decision loop. The top half of Fig. 6.3 (in blue) shows prediction enabled through reflection that can be utilized in the decision-making process of a feedback loop. The main goal of the predictive model is to estimate system behavior based on potential actuation decisions as well as system dynamics.

6.2.1   Middleware for Reflective Decision-Making

The increasing heterogeneity in a platform's resource types and the interactions between resources pose challenges for coordinated model-based decision-making in the face of dynamic workloads. Self-awareness properties address these challenges for emerging SoC platforms through reflective resource managers. Reflective resource managers build a model of the system which represents the software organization or the architecture of the target platform. Resource managers can use reflective models to anticipate the effects of changing the system configuration

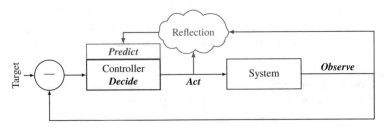

Fig. 6.3 Feedback loop overview. The bottom part of the figure represents a simple observe–decide–act loop. The top part (in blue) adds the reflection mechanism to this loop, enabling predictions for smart decision-making

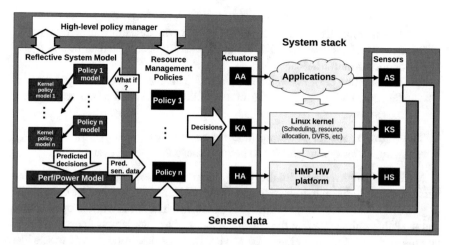

Fig. 6.4 MARS framework overview from [10]. Different layers of the system stack coordinate through policies to orchestrate the management of resources: sensors inform policies of the system state; policies coordinate with models to perform reflective queries, and make resource management decisions; policies set actuators to enact changes on the system

at runtime. However, with SoC computing platform architectures evolving rapidly, porting the self-aware decision logic across different hardware platforms is challenging, requiring resource managers to update their models and platform-specific interfaces. To address this problem, we propose MARS (Middleware for Adaptive and Reflective Systems), a cross-layer and multi-platform framework that allows users to easily create resource managers by composing system models and resource management policies in a flexible and coordinated manner.

Figure 6.4 shows an overview of the MARS framework (shaded), with *Sensors* and *Actuators* interfacing across multiple layers of the system stack: *Applications*, *Linux kernel*, and *HW Platform*. The components of MARS are explained next.

1. **Sensors and actuators**: The sensed data consists of performance counters (e.g., instructions executed, cache misses, etc.) and other sensory information (e.g., power, temperature, etc.). The collected data is used to assess the current system state and to characterize workloads. Any updates to the system configuration (e.g., CPU core frequency, GPU frequency, memory controller frequency, task-to-core mapping) happen through system knobs. Actuators allow system configuration changes to optimize operating point or control trade-offs.
2. **Resource Management Policies**: They are platform agnostic user-level daemons implemented in MARS using supported sensors, actuators, and reflective system models.
3. **Reflective system model** is used by the policies to make informed decisions. The reflective model has the following subcomponents:

(a) Models of *policies implemented by the underlying OS kernel* used for coordinating decisions made within MARS with decisions made by the OS.
(b) Models of *user policies* that are automatically instantiated from any policy defined within MARS.
(c) The baseline *performance/power model*. This model takes as input the predicted actuations generated from the policy models and produces predicted sensed data.

4. **The policy manager** is responsible for reconfiguring the system by adding, removing, or swapping policies to better achieve the current system goal.

MARS is implemented in the C++ language following an object-oriented paradigm and works on hardware (e.g., Odroid-XU3, Nvidia Jetson TX2), simulated (e.g., gem5), and trace-based offline [11] platforms. The framework is open source and available online.[2] While the current version of MARS targets energy-efficient heterogeneous SoCs, we believe the MARS framework can be ported to a wider range of systems (e.g., webservers, high-performance clusters) to support self-aware resource management.

6.3 Managing Energy-Efficient Chip Multiprocessors

Dynamic resource management for HMPs is a well-known challenge: integration of hundreds of cores running various workloads with conflicting constraints increases the pressure on limited shared system resources. A promising and well-established approach is the use of control-theoretic solutions based on rigorous mathematical formalisms that can provide bounds and guarantees for system resource management. In this context, we discuss efforts that deploy control-theoretic-centric runtime resource management of HMPs, from simple Single Input Single Output (SISO) controllers to more complex Supervisory Control Theory (SCT) methods.

6.3.1 Single Input Single Output Controllers

Conventional control theory methods proposed for resource management use Single Input Single Output (SISO) controllers for the ease in deployment and the guarantees they provide in tracking the target output. These SISO controllers use Proportional Integral (PI), Proportional Integral Derivative (PID), or lead-lag implementations [22]. Figure 6.5 depicts a first-order feedback SISO controller which can be deployed either as a PI or a PID controller. The error e is the input to the controller. Note that to compute the current control input u, the controller

---

[2]Code repository at https://github.com/duttresearchgroup/MARS.

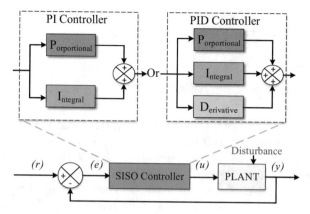

Fig. 6.5 *Single Input Single Output* (SISO) feedback loop

needs to have the current value of the error e along with the past value of the error and the past value of the control input. It is this memory inherent in the controller that makes it dynamic.

6.3.2 Multiple Input Multiple Output Controllers

Modern HMPs execute diverse set of workloads with varying resource demands, which sometimes exhibit conflicting constraints. In this context, the use of SISO controllers might not be effective as multiple system goals varying over time need to be managed in a coordinated and holistic manner. Multiple Input Multiple Output (MIMO) control theory is able to coordinate and prioritize multiple design goals and actions. MIMO controllers have proven effective for coordinating management of multiple goals in unicore processors [17] and HMPs [12].

6.3.3 Adaptive Control Methods

Ideally, control-theoretic solutions should provide formal guarantees, be simple enough for runtime implementation, and handle nonlinear system behavior. Static linear feedback controllers such as SISO and MIMO can provide robustness and stability guarantees with simple implementations, while adaptive controllers modify the control law at runtime to adapt to the discrepancies between the expected and the actual system behavior. However, modifying the controller at runtime is a costly operation that also invalidates the formal guarantees provided at design time. In order to be able to take predicted responsive actions against nonlinear behavior of the computer systems, a well-established and lightweight adaptive control-theoretic technique called *Gain Scheduling* can be used. This method is used for dynamic power management in chip multiprocessors in [2].

6.3.4 Hierarchical Controllers

Supervisory Control Theory (SCT) [19, 30] provides formal and systematic supervision of classical MIMO/SISO controllers. SCT uses modular decomposition of control problems to manage their complexity. Specifically, supervisory control has two key properties: (1) rapid adaptation in response to abrupt changes in management policy and (2) low computational complexity by computing control parameters for different policies offline. New policies and their corresponding parameters can be added to the supervisor on demand. Therefore, SCT is suitable for resource management problems (such as managing power, energy, and quality-of-service metrics) that can be modeled using logic and discrete system dynamics.

Figure 6.6 depicts a high-level view of supervisory control for HMP resource management. Either the user or the system software may specify *Variable Goals and Policies*. The *Supervisory Controller* aims to meet system goals by managing the low-level controllers. High-level decisions are made based on the feedback given by the *High-level Plant Model*, which provides an abstraction of the entire system. Various types of *Classic Controllers*, such as PID or state-space controllers, can be used to implement each low-level controller based on the target of each subsystem. The flexibility to incorporate any pre-verified off-the-shelf controllers without the need for system-wide verification is essential for the modularity of this approach. The supervisor provides parameters such as output references or gain values to each low-level controller during runtime according to the system policy. Low-level controller subsystems update the high-level model to maintain global system state,

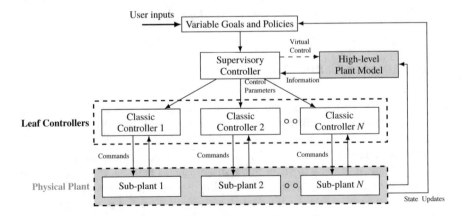

Fig. 6.6 High-level view of Supervisory Control Theory

and potentially trigger the supervisory controller to take action. The high-level model can be designed in various fashions (e.g., rule-based or estimator-based) to track the system state and provide the supervisor with guidelines. Supervisory control provides the opportunity to benefit from both classical control-theoretic methods and heuristics in a robust fashion. The SCT hierarchy in Fig. 6.6 is successfully used to manage quality-of-service (QoS) goals within a power budget on an HMP in [18].

6.4 Heterogeneous Mobile Governors: Energy-Efficient Mobile System-on-a-Chip

Mobile games stress modern SoCs by utilizing heterogeneous processing elements, CPUs and GPUs, concurrently. However, the utilization of each processing element may vary between games. Performance of these games that usually is measured in frames per second (FPS) can highly depend on the operating frequency of compute units. However, conventional DVFS governors conservatively choose high frequencies without considering the utilization pattern of the games [16]. In order to meet a performance goal while conserving energy, the frequency of each processing element should be as low as possible without an observable effect on the FPS.

6.4.1 Sensors to Capture Dynamism

To coordinate frequency configuration decisions, a cooperative CPU-GPU DVFS strategy, Co-Cap [14], limits the maximum frequency of CPUs and GPUs on a game-specific basis. Based on the utilization of each processing element, games are classified as one of the following classes: (1) No CPU-GPU Dominant; (2) CPU Dominant; (3) GPU Dominant; and (4) CPU-GPU Dominant. Figure 6.7 shows the classes and gives an example of each class. To determine a maximum frequency for each game class, Co-Cap implements a frame rate sensor, which is affected by both CPU and GPU frequencies. By limiting maximum frequencies for each game class, Co-Cap reduces energy consumption without observable performance degradation.

The assumption in Co-Cap is that games can only belong to one of the classes. However, some games might change their dynamic behavior throughout their life cycle. To proactively respond to the dynamic CPU and GPU frequency requirements of games, a DVFS governor policy requires more information about a game's workload dynamism. A Hierarchical Finite State Machine (HFSM) based CPU-GPU governor, HiCAP [13], models the dynamic behavior of mobile gaming workloads and applies a cooperative, dynamic CPU-GPU frequency-capping policy to conserve energy by adapting to a game's inherent dynamism. Using the HFSM, a DVFS governor can predict the next workload feature for a certain window

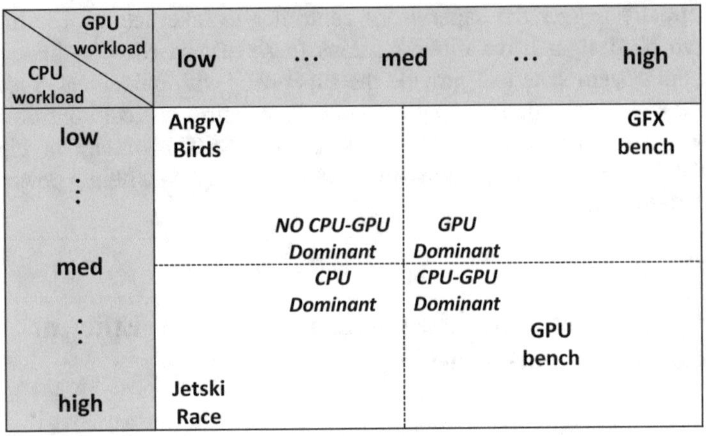

GPU workload CPU workload	low	...	med	...	high
low	Angry Birds				GFX bench
:					
med		*NO CPU-GPU Dominant*	*GPU Dominant*		
		CPU Dominant	*CPU-GPU Dominant*	GPU bench	
:					
high	Jetski Race				

Fig. 6.7 Classification of CPU-GPU workloads observed in mobile games from [14]

at a game's runtime. Through this added self-awareness, HiCAP reduces energy consumption even further than Co-Cap.

Further dynamism exists in a game's memory access patterns. Some scenes in mobile games read more graphics data than others, resulting in increased memory utilization. This may slow down the CPU portion of the game, but on the other hand when memory utilization is low, it may run faster than originally predicted by a conventional DVFS governor. A conventional DVFS governor cannot detect these memory utilization changes by sensing utilization, causing prediction errors to increase. MEMCOP, a Memory-aware Cooperative Power Management Governor for Mobile games [5], senses the number of last level cache misses to monitor the memory pressure of the system in addition to CPU, and GPU memory utilization. This prevents the CPU DVFS governor from increasing frequency due to inaccurate predictions caused by variation in memory access time.

6.4.2   Toward Self-Aware Governors

Co-Cap, HiCap, and MEMCOP DVFS policies are each steps toward a self-aware DVFS governor policy for heterogeneous SoCs. Each policy monitors system's state using novel sensors, and defines runtime prediction rules to reflect and adapt to changes in mobile game behavior. However, the predictive models are generated statistically at design time, and remain the same during the execution. Moreover, as the predictive model becomes more complex, prediction errors increase due to the assumption of a linear relationship between the model's input and output. ML-Gov, a machine learning enhanced integrated CPU-GPU governor [15], tries to address these issues by applying machine learning algorithms. This method does not require rule tuning at design-time. ML-Gov's machine learning algorithm helps to exploit nonlinear characteristics between frequency and performance. ML-Gov currently builds the model offline, but through enhanced self-awareness via online updates of the reflective model, could adapt to previously unknown games and classes.

6.5 Adaptive Memory: Managing Runtime Variability

Heterogeneous processing elements on mobile SoCs share limited memory resources, leading to memory contention and stalled processes waiting for data. This performance degradation is exacerbated by the Von Neumann bottleneck, a prevalent problem in modern day computer systems. Data transfer speeds in memory have not been able to keep up with the performance gains of processors exemplified by Moore's law. However, with the end of Moore's law on the horizon there is an ever increasing need to alleviate the Von Neumann bottleneck to increase the performance of computer systems. There have been various approaches over the years to address the Von Neumann bottleneck such as putting critical memory in an easily accessible cache [25] and recently in an easily accessible Software-Programmable Memory (SPM), also known as a scratchpad, using multi-threading [9], and exploiting cache-coherency [7]. We address the bottleneck by providing self-awareness with respect to memory resource utilization.

6.5.1 Sharing Distributed Memory Space

Software-Programmable Memories are a promising alternative to hardware-managed caches in embedded systems. However, traditional approaches for managing SPMs do not support sharing of distributed memory resources, missing the opportunity to utilize those memory resources. Employing operating-system-level awareness of SPM utilization, memory resources can be shared by allowing threads to opportunistically exploit the entire memory space for unpredictable application workloads. Best-effort policies can be used to maximize the usage of on-chip SPMs. The policies can be supported by hardware via distributed memory management units (MMUs), an on-chip component that can be used to exchange information between the NoC and an MMU's local SPM. Sharing distributed SPM space reduces memory contention, resulting in reduced memory latency by reducing off-chip memory accesses by about 14%. The off-chip access reduction decreases average execution time by about 19.5%, which in turn reduces energy consumption [23, 29]. More intelligent policies that explore a mixed SPM/cache hierarchy for many-core embedded systems can yield further improvements.

6.5.2 Memory Phase Awareness

Modern mobile devices use multi-core platforms that allow for concurrent execution of multimedia and non-multimedia applications that enter and exit at unpredictable times. Each application also has variable memory demands during these unpredictable times. By being aware of the periodic patterns, or *phasic behavior*, of an application's memory usage (memory phases), a system's on-chip memory can be more efficiency utilized. Memory phases can be identified from memory usage information extracted on an application basis, and can be used to prioritize different memory pages in a multi-core platform without having any prior knowledge about running applications. The identification process can be integrated into the runtime system and done online. For example, memory phases can be used for effective sharing of distributed SPMs for multi-core platforms to reduce memory access latency and contention. Experiments on workloads with varying intra- and inter-application memory-intensity show that using phase detection schemes can reduce memory access latency up to 45% for configurations up to 16 cores [28]. Ongoing work investigates more aggressive use of memory phasic behavior in many-core architectures with hundreds of cores.

6.5.3 Quality-Configurable Memory

We have established how self-awareness can be achieved through formal control theory. Figure 6.8 shows a closed feedback control loop with a quality monitor that can measure memory utilization and processor usage with respect to a QoS goal to fit the runtime requirements of applications. The quality monitor gives a quality score and sends the collected data to a high-level controller. The controller reflects on the data, then tunes knobs to adapt the memory utilization and processor usage to minimize the error between the current quality and the quality-of-service goal. The self-aware approach enables dynamic convergence toward dynamic memory utilization and quality targets for unpredictable workloads. While current

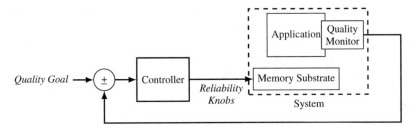

Fig. 6.8 Control loop implementing a self-aware approach to memory utilization optimization. The controller optimizes memory knobs to improve application performance

results indicate that a self-aware memory controller outperforms a manual quality configuration scheme, there is much work to be done with to analyze energy trade-offs when using a self-aware memory controller, and whether a MIMO controller could be more effective for resource management in many-core systems with the self-aware approach [24].

6.6 What's Ahead?

Self-awareness enables a system to observe its context and make changes to optimize its execution at runtime. For instance, it is possible to allow a system to tune its execution to optimize power consumption. Through observing how it has reacted to past changes in certain conditions, the system can learn what the impact on the overall execution and power consumption was, and if a different adaptation would be more appropriate in the future. To further explore such opportunities in computing systems, we shift our focus to a new project: the Information Processing Factory (IPF). IPF is a step toward autonomous many-core platforms in cyber-physical systems (CPS) and the Internet of Things (IoT). It represents a paradigm shift in platform design, with robust and independent platform operation in the focus of platform-centric design rather than existing semiconductor device or software technology, as mostly seen today [4].

We use the metaphor of an Information Processing Factory to draw similarities between microelectronics systems and factories as follows: in a factory, all components must adapt to the current workload [20]. Additionally, this adaptation cannot be done offline and must instead be done in real time without interrupting the baseline operations. Future microelectronic systems (e.g., MPSoCs) should operate in a similar manner.

Clusters of component-specific, uncorrelated control occurrences cannot handle operations of large scale systems with multi-criteria objective functions. Similarly, a centralized controller model is also inadequate in this case because it cannot scale. The goal of the IPF project is to demonstrate that a hybrid hierarchical approach, sporting as much modularity as possible and as much centralized as necessary, is a much more effective means of achieving the desired goal while maintaining cost efficiency, low overhead, and scalability.

Figure 6.9 depicts how we envision the platform to be structured. Information provided by sensors is gathered and merged into self-organizing, self-aware (SO/SA) control processing instances across different hardware/software abstraction layers comprising an MPSoC-based CPS system. The SO/SA instances generate actuation directives affecting the MPSoC system components at same or lower levels of abstraction. The SO/SA paradigm is not limited in scope to optimization of CPS operational parameters/metrics. In fact, self- and group-awareness can also enable higher level tasks such as self-protection of both the MPSoC and the overall CPS system.

Fig. 6.9 Idealized IPF
Platform from [20].
Self-organizing and
Self-aware (SO/SA)
components are distributed
throughout the system stack.
Both hardware and
application system state are
monitored and configurations
adjusted using sensors and
actuators

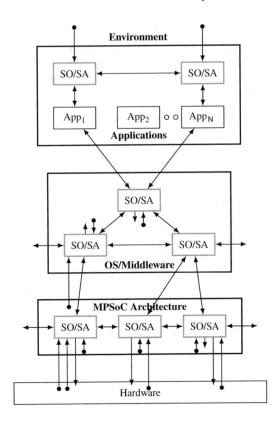

6.6.1   Example Use Case: Autonomous Driving

The key innovation in automated driving as compared to driver assistance systems
is the transition of decision-making from the driver to the vehicle. The application
processing and communication requirements ask for platform performance, memory
capacity, and communication bandwidth and latency far beyond the capabilities of
current architectures. At the same time, these platforms must be highly reliable
and guarantee sufficient functionality under platform errors, aging, and degradation
to meet safety standards. That is, platforms and their components must be fail-
operational, i.e., must be able to continue driving, instead of fail-safe, as today.

Thus, the automated driving requirements can be mapped to corresponding
requirements of an Information Processing Factory. The system must be capable of
in-field integration, i.e., able to adapt to changes in the workload of both critical and
non-critical (best-effort) functions. The system must find a new suitable mapping
and must prevent the changes from violating the guarantees of other software
components. The software must be able to detect and to adapt to transient errors
in order to provide a reliable service. This requires self- diagnosis and self-healing.

The system must be predictable and provide for minimum performance guarantees for all scenarios.

Allowing and exploiting dynamic system behavior through IPF can significantly improve platform performance and resource utilization. Thus, the system must be able to optimize the execution and mapping online: self-optimization. The optimization may target e.g. aging (temperature), power consumption, response time, and resource utilization.

6.7 Summary

Future cyber-physical systems will host a large number of coexisting distributed applications on hardware platforms with thousands to millions of networked components communicating over open networks. These distributed applications will include both critical and best-effort tasks, may be subject to permanent change, environment dynamics and application interference. Using wisdom gathered from our initial exploration into self-aware SoCs, we introduce a new Information Processing Factory paradigm to manage current and future cyber-physical systems.

Acknowledgments The authors would like to acknowledge Santanu Sarma, Chen-Ying Hsieh, JurnGyu Park, Majid Shoushtari, and Hossein Tajik for their research contributions. We acknowledge financial support by NSF grant CCF-1704859, and the Marie Curie Actions of the European Union's H2020 Program.

References

1. B. Donyanavard, T. Mück, S. Sarma, N. Dutt, Sparta: runtime task allocation for energy efficient heterogeneous many-cores, in *Proceedings of the Eleventh IEEE/ACM/IFIP International Conference on Hardware/Software Codesign and System Synthesis, CODES '16* (ACM, New York, 2016), pp. 27:1–27:10. https://doi.org/10.1145/2968456.2968459
2. B. Donyanavard, A.M. Rahmani, T. Muck, K. Moazemmi, N. Dutt, Gain scheduled control for nonlinear power management in CMPs, in *2018 Design, Automation Test in Europe Conference Exhibition (DATE)* (2018)
3. N. Dutt, A. Jantsch, S. Sarma, Self-aware cyber-physical systems-on-chip, in *2015 IEEE/ACM International Conference on Computer-Aided Design (ICCAD)* (2015), pp. 46–50. https://doi.org/10.1109/ICCAD.2015.7372548
4. N. Dutt, F.J. Kurdahi, R. Ernst, A. Herkersdorf, Conquering MPSoC complexity with principles of a self-aware information processing factory, in *Proceedings of the Eleventh IEEE/ACM/IFIP International Conference on Hardware/Software Codesign and System Synthesis* (ACM, New York, 2016), p. 37
5. C.Y. Hsieh, J.G. Park, N. Dutt, S.S. Lim, MEMCOP: memory-aware co-operative power management governor for mobile games. Des. Autom. Embed. Syst. **22**(1-2), 95–116 (2018). https://doi.org/10.1007/s10617-018-9201-8
6. A. Jantsch, N. Dutt, A.M. Rahmani, Self-awareness in systems on chip– a survey. IEEE Des. Test **34**(6), 8–26 (2017). https://doi.org/10.1109/MDAT.2017.2757143

7. O. Kayiran, N.C. Nachiappan, A. Jog, R. Ausavarungnirun, M.T. Kandemir, G.H. Loh, O. Mutlu, C.R. Das, Managing GPU concurrency in heterogeneous architectures, in *2014 47th Annual IEEE/ACM International Symposium on Microarchitecture* (2014), pp. 114–126. https://doi.org/10.1109/MICRO.2014.62

8. S. Kounev, P. Lewis, K.L. Bellman, N. Bencomo, J. Camara, A. Diaconescu, L. Esterle, K. Geihs, H. Giese, S. Götz, P. Inverardi, J.O. Kephart, A. Zisman, *The Notion of Self-aware Computing* (Springer International Publishing, Cham, 2017)

9. C.-K. Luk, Tolerating memory latency through software-controlled pre-execution in simultaneous multithreading processors, in *Proceedings 28th Annual International Symposium on Computer Architecture* (2001), pp. 40–51. https://doi.org/10.1109/ISCA.2001.937430

10. T.R. Mück, Reflective on-chip resource management policies for energy-efficient heterogeneous multiprocessors. Ph.D. Thesis, University of California, Irvine (2018). http://www.escholarship.org/uc/item/6n93v21h

11. T. Mück, B. Donyanavard, N. Dutt, PoliCym: rapid prototyping of resource management policies for HMPs, in *Proceedings of the 28th International Symposium on Rapid System Prototyping: Shortening the Path from Specification to Prototype, RSP '17* (ACM, New York, 2017), pp. 23–29. https://doi.org/10.1145/3130265.3130321

12. T.R. Muck, B. Donyanavard, K. Moazzemi, A.M. Rahmani, A. Jantsch, N.D. Dutt, Design methodology for responsive and robust MIMO control of heterogeneous multicores. IEEE Trans. Multi Scale Comput. Syst. **4**, 944–951 (2018). https://doi.org/10.1109/TMSCS.2018.2808524

13. J.G. Park, N. Dutt, H. Kim, S.S. Lim, HiCAP: hierarchical FSM-based dynamic integrated CPU-GPU frequency capping governor for energy-efficient mobile gaming, in *ISLPED '16: Proceedings of the 2016 International Symposium on Low Power Electronics and Design* (2016), pp. 218–223. https://doi.org/10.1145/2934583.2934588

14. J.G. Park, C.Y. Hsieh, N. Dutt, S.S. Lim, Co-cap: energy-efficient cooperative CPU-GPU frequency capping for mobile games, in *SAC '16: Proceedings of the 31st Annual ACM Symposium on Applied Computing* (ACM, New York, 2016). https://doi.org/10.1145/2851613.2851671

15. J.G. Park, N. Dutt, S.S. Lim, ML-Gov: a machine learning enhanced integrated CPU-GPU dVFS governor for mobile gaming, in *ESTIMedia '17: Proceedings of the 15th IEEE/ACM Symposium on Embedded Systems for Real-Time Multimedia* (2017), pp. 12–21. https://doi.org/10.1145/3139315.3139317

16. J.G. Park, C.Y. Hsieh, N. Dutt, S.S. Lim, Synergistic CPU-GPU frequency capping for energy-efficient mobile games. ACM Trans. Embed. Comput. Syst. **17**(2), 1–24 (2017). https://doi.org/10.1145/3145337

17. R.P. Pothukuchi, A. Ansari, P. Voulgaris, J. Torrellas, Using multiple input, multiple output formal control to maximize resource efficiency in architectures, in *Proceedings of the 43rd International Symposium on Computer Architecture* (2016)

18. A.M. Rahmani, B. Donyanavard, T. Müch, K. Moazzemi, A. Jantsch, O. Mutlu, N. Dutt, Spectr: formal supervisory control and coordination for many-core systems resource management, in *Proceedings of the Twenty-Third International Conference on Architectural Support for Programming Languages and Operating Systems, ASPLOS '18* (ACM, New York, 2018), pp. 169–183. https://doi.org/10.1145/3173162.3173199

19. P.J. Ramadge, W.M. Wonham, The control of discrete event systems. Proc. IEEE **77**(1), 81–98 (1989)

20. A. Sadighi, B. Donyanavard, T. Kadeed, K. Moazzemi, T. Mück, A. Nassar, A.M. Rahmani, T. Wild, N. Dutt, R. Ernst, et al., Design methodologies for enabling self-awareness in autonomous systems, in *2018 Design, Automation & Test in Europe Conference & Exhibition (DATE)* (IEEE, Piscataway, 2018), pp. 1532–1537

21. S. Sarma, N. Dutt, P. Gupta, A. Nicolau, N. Venkatasubramanian, On-chip self-awareness using cyberphysical-systems-on-chip (CPSoC), in *Proceedings of the 2014 International Conference on Hardware/Software Codesign and System Synthesis, CODES '14* (ACM, New York, 2014), pp. 22:1–22:3. https://doi.org/10.1145/2656075.2661648

22. S. Shahosseini, K. Moazzemi, A.M. Rahmani, N. Dutt, Dependability evaluation of SISO control-theoretic power managers for processor architectures, in *2017 IEEE Nordic Circuits and Systems Conference (NORCAS): NORCHIP and International Symposium of System-on-Chip (SoC)* (2017)

23. M. Shoushtari, B. Donyanavard, L.A.D. Bathen, N. Dutt, Shave-ice: sharing distributed virtualized SPMs in many-core embedded systems. ACM Trans. Embed. Comput. Syst. **17**(2), 47:1–47:25 (2018). https://doi.org/10.1145/3157667

24. M. Shoushtari, A. Rahmani, N. Dutt, Quality-configurable approximate memory hierarchy: a formal control theory approach. Workshop on Approximate Computing Across the Stack (2018). http://approximate.computer/wax2018/papers/wax2018-paper12.pdf

25. A.J. Smith, Cache memories. ACM Comput. Surv. **14**(3), 473–530 (1982). https://doi.org/10.1145/356887.356892

26. B.C. Smith, *Reflection and Semantics in a Procedural Programming Language*. Ph.D. (MIT Press, Cambridge, 1982)

27. V. Spiliopoulos, S. Kaxiras, G. Keramidas, Green governors: a framework for continuously adaptive DVFS, in *Proceedings of the 2011 International Green Computing Conference and Workshops, IGCC '11* (IEEE Computer Society, Washington, 2011), pp. 1–8. https://doi.org/10.1109/IGCC.2011.6008552

28. H. Tajik, B. Donyanavard, N. Dutt, On detecting and using memory phases in multimedia systems, in *Proceedings of the 14th ACM/IEEE Symposium on Embedded Systems for Real-Time Multimedia, ESTIMedia'16* (ACM, New York, 2016), pp. 57–66. https://doi.org/10.1145/2993452.2993566

29. H. Tajik, B. Donyanavard, N. Dutt, J. Jahn, J. Henkel, Spmpool: runtime SPM management for memory-intensive applications in embedded many-cores. ACM Trans. Embed. Comput. Syst. **16**(1), 25:1–25:27 (2016). https://doi.org/10.1145/2968447

30. J. Thistle, Supervisory control of discrete event systems. Math. Comput. Model. **23**(11), 25–53 (1996)

Chapter 7
Pushing the Limits of Parallel Discrete Event Simulation for SystemC

Rainer Dömer, Zhongqi Cheng, Daniel Mendoza, and Emad Arasteh

7.1 Introduction

The IEEE standard SystemC language [13] is widely used for the specification, modeling, validation, and evaluation of electronic system level (ESL) models. The Accellera Systems Initiative maintains not only the official SystemC language definition, but also provides an open source proof-of-concept library that can be used to simulate SystemC design models [1]. However, implementing the classic scheme of discrete event simulation (DES), this reference simulator runs sequentially and cannot utilize the parallel computing resources available on multi- and many-core processor hosts. This severely limits the execution speed of SystemC simulation.

In order to provide faster execution, parallel discrete event simulation (PDES) [8, 12] techniques can be applied. While significant obstacles exist specifically for the SystemC language [7], many parallel simulation approaches have been proposed [5, 11, 19, 21–24]. Beyond these synchronous PDES techniques, *out-of-order* PDES [6] is even more aggressive. By localizing the simulation time to individual threads and carefully handling events at different times, the simulator engine can issue threads in parallel and *ahead of time*, following a partial ordering without loss of accuracy. This results in better exploitation of the available parallelism and thus maximum simulation speed.

The *Recoding Infrastructure for SystemC (RISC)* project described in this paper implements out-of-order PDES for the IEEE SystemC language as open source. Specifically, RISC provides a dedicated SystemC compiler and corresponding out-of-order parallel simulator [2, 8, 16]. Compared to the other approaches, RISC automatically analyzes the SystemC source code, identifies all potential race condi-

R. Dömer (✉) · Z. Cheng · D. Mendoza · E. Arasteh
Center for Embedded and Cyber-Physical Systems, University of California, Irvine, CA, USA
e-mail: doemer@uci.edu

© The Author(s) 2021
J.-J. Chen (ed.), *A Journey of Embedded and Cyber-Physical Systems*,
https://doi.org/10.1007/978-3-030-47487-4_7

tions, and then instruments the model to prevent any conflicts. This transformation does not require any manual recoding or application-specific knowledge.

We share our RISC proof-of-concept implementation with the EDA community as an open source software project in order to facilitate evaluation, promote parallel SystemC simulation, and achieve fruitful collaboration [3, 4].

7.2   RISC Framework

While the RISC software framework may be used for many other analysis and transformation tasks on SystemC models, parallel simulation is the main purpose. To perform semantics-compliant parallel simulation with out-of-order scheduling, we introduce a dedicated SystemC compiler that works hand in hand with a new simulator. This is in contrast to the traditional SystemC simulation flow where a SystemC-agnostic C++ compiler includes the SystemC headers and links the design model directly against the Accellera reference library.

As shown in Fig. 7.1, the RISC compiler acts as a frontend that processes the input model and generates an intermediate model with special instrumentation for conflict-free parallel execution. The instrumented model is then linked against the extended RISC SystemC library by the target compiler (a regular C++ compiler, such as GNU gcc or Intel icpc) in order to produce the output executable model. Out-of-order parallel simulation is then performed simply by running the generated executable model.

From the user perspective, we simply replace the regular C++ compiler with the SystemC-aware RISC compiler (which in turn calls the underlying C++ compiler). Otherwise, the overall SystemC validation flow remains the same as the traditional tool flow. Simulation is just faster due to the parallel execution. Note also that this process is fully automated. No user interaction or manual code transformation is necessary.

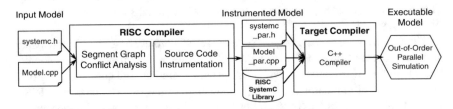

Fig. 7.1   RISC tool flow for out-of-order parallel simulation of SystemC models [16]

7.2.1 RISC Compiler

In order to produce a safe parallel model, the RISC compiler performs three major tasks, namely segment graph construction, conflict analysis, and finally source code instrumentation.

7.2.1.1 Segment Graph Construction

A segment graph (SG) [6] is a directed graph that represents the source code segments executed during the simulation between scheduling steps. More specifically, every segment is associated with a corresponding scheduler entry point, namely a `wait` statement in SystemC. All other statements in the SystemC source code become part of those segment nodes where they are executed when the `wait` statement resumes its execution.

The segment graph construction is a fully automatic but complex process which we will not describe here (see [6] for detailed coverage). However, the RISC compiler must parse the SystemC input model first into an Abstract Syntax Tree (AST). Since SystemC is a syntactically regular C++ code, RISC relies here on the ROSE compiler infrastructure [18]. The ROSE internal representation (IR) provides RISC with a powerful C/C++ compiler foundation that supports AST generation, traversal, analysis, and transformation.

As illustrated with the RISC software stack shown in Fig. 7.2, the RISC compiler then builds a SystemC IR on top of the ROSE IR which accurately reflects the SystemC structures, including the module and channel hierarchy, port connectivity, and other SystemC-specific constructs. On top of the SystemC IR, the compiler architecture then builds the Segment Graph generator and data structures, as well as all other RISC analysis and transformation functions.

7.2.1.2 Conflict Analysis

The segment graph data structure serves as the foundation for segment conflict analysis. At run time, the scheduler in the simulator must ensure that every parallel thread to be issued has no conflicts with any other threads currently in the $READY$

Fig. 7.2 Software stack of the RISC compiler [8]

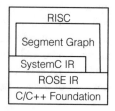

and *RUN* queues. For this we use the RISC compiler to detect any possible conflicts between these threads already at compile time.

Potential conflicts in SystemC include data hazards, event hazards, and timing hazards, all of which may exist among the segments executed by the threads considered for parallel execution. Again, we refer to [6] for a detailed discussion of these hazards and their static or dynamic detection in RISC. However, we note that if the hazards would be ignored, this would lead to race conditions at run time and jeopardize the correctness of the SystemC simulation.

7.2.1.3   Source Code Instrumentation

As a result of the conflict analysis, the RISC compiler generates a set of conflict and timing tables that describe all possible hazards between any two threads. Using this conservative conflict information, the simulator can then at run time quickly determine by a simple table look-up whether or not it is safe to issue a given thread in parallel or ahead of time.

As shown above in Fig. 7.1, the RISC compiler and simulator work closely together. The compiler performs conservative conflict analysis and passes the analysis results to the simulator which then can make safe scheduling decisions quickly.

To pass information from the compiler to the simulator, we use automatic source code instrumentation. That is, the intermediate model generated by the compiler contains instrumented (automatically generated) code which the simulator can then safely rely on.

At the same time, the RISC compiler also instruments the SystemC `wait` statements with corresponding segment ID and furnishes user-defined channels with automatic protection against race conditions among communicating threads.

7.2.2   RISC Simulator

The RISC simulator supports out-of-order discrete event simulation (OoO PDES) [6] for fast SystemC simulation. In OoO PDES, we break the strict order of time (the synchronous barrier) by localizing time stamps to each thread. Since each thread has its own time stamp, the OoO PDES scheduler relaxes the event and simulation time updates, allowing more threads (at different simulation cycles) to run in parallel and ahead of time. This results in a higher degree of parallelism and thus higher simulation speed. We are using advanced static compile-time analysis to identify all such potential conflicts. Based on this information (a simple table look-up is sufficient), the OoO PDES scheduler can then at run time quickly decide whether or not a set of threads has any conflicts with each other.

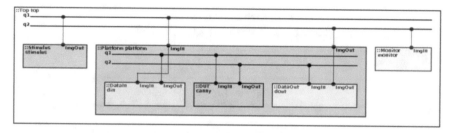

Fig. 7.3 Module hierarchy visualization of a SystemC model of a Canny edge detector [17]

7.2.3   RISC Analysis and Transformation Tools

As an example of other SystemC analysis tools built on top of RISC, `visual` [17] enables the user to visualize the SystemC module hierarchy. It supports a graphical user interface implemented with the Gtk API and renders a specified SystemC source file's module hierarchy, which is drawn using the Cairo API. The tool obtains module data from the SystemC IR in the RISC software stack which contains information about nested modules and thus can recursively iterate through nested lists of child modules in order to obtain enough information to visualize the hierarchy of the entire SystemC source file. The input SystemC source file may contain thousands of lines of code which can make manually drawing a representation of the modules, ports, and channels described by the code a difficult and time-consuming task. Thus the `visual` tool was created to address this issue. It can automatically generate a visual representation of a SystemC model in a very short period of time. Figure 7.3 shows the module visualization of a Canny edge detector application.

7.3   Experiments

We will now evaluate the performance of the RISC simulator. The following experiments show the speedup on an Intel Xeon Phi™ Coprocessor 5110P many-core architecture. The coprocessor contains 60 cores where each core has a vectorization unit of 512 bit. To obtain unambiguous measurements, we turn CPU frequency scaling off for all experiments.

7.3.1   Mandelbrot Renderer

The Mandelbrot renderer is a parallel video application to compute the Mandelbrot set. Basically, the device under test (DUT) hosts a number of renderer units. Each

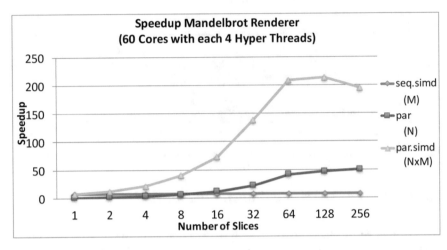

Fig. 7.4 Speedup of the Mandelbrot Renderer [20]

unit computes a different slice of the Mandelbrot image. At compile time, the user defines how many slices are available.

Figure 7.4 shows the simulation results [20]. Due to the minimal communication needs in this application, highest speedups are reached. The vectorization unit with 512 bit can execute up to eight double-precision floating-point operations in parallel. A speedup M of 6.9x is achieved. The thread-level parallelization increases strongly on the 60 cores with a speedup N of 50x. Afterwards, the speed slows down due to the 60 physical cores and use of hyper-threads. Notably, the combination of the thread and data level parallelization $N \times M$ generates a speedup of up to 212x.

7.4   RISC Open Source Project

We make the Recoding Infrastructure for SystemC (RISC) described in this article freely available online as a software artifact [9]. Generally, an artifact is a software program together with an applicable data set and test suite that accompanies a research publication for the purpose of independent evaluation.[1]The point here is that the proposed algorithms and data structures are made available as proof-of-concept implementation and can be used and evaluated by others. Experimental results may be replicated and validated. The proposed approach can also be compared against related work and in the presence of source code even be extended. Otherwise, great challenges are posed in repeatability [15].

---

[1]Because of its importance, artifact evaluation has been adopted as integral part of the review process in several computer science areas, such as Software Engineering and Programming Languages [10, 14].

Specifically, the presented RISC compiler and simulator are available as open source on the web [2] and can be used without restrictions (BSD license terms). RISC can be downloaded in both source code and binary format.

7.4.1   Open Source Code and Documentation

In its current version [4], the RISC open source package consists of approximately 162,000 lines of code and includes the C++ source code for the RISC compiler and simulator, Linux build scripts and installation instructions, as well as comprehensive documentation of the compiler and simulator APIs and tool manual pages. Example SystemC models, such as an abstract DVD player and the Mandelbrot renderer, and a regression test bench are included as well.

Given a suitable Linux platform,[2] the RISC source code package can be easily installed and then tested. After downloading and adjusting the installation Makefile, a simple make all command builds and installs the RISC framework and runs several demo examples. The user can then fully evaluate the software with other SystemC examples and even extend our proof-of-concept implementation with new features.

7.4.2   Binary Image for "Plug-and-Play" Evaluation

For a quick test run without compilation and installation, we also provide a Docker container [3] for using RISC in "plug-and-play" fashion. The Docker image contains RISC (and all needed libraries) in binary format and allows the user to test it with just a few Linux commands, as shown in Fig. 7.5.

```
bash# docker pull ucirvinelecs/risc
bash# docker run -it ucirvinelecs/risc
[dockeruser]# cd demodir
[dockeruser]# make test
```

Fig. 7.5 Linux commands to use RISC in a Docker container

---

[2]Red Hat Enterprise and CentOS Linux version 6 and 7 are verified to work.

7.5   Conclusion

The Recoding Infrastructure for SystemC (RISC) provides an automatic compiler-based framework to analyze and simulate IEEE SystemC models in parallel. In particular, we have introduced the RISC compiler and simulator. Using automatic conflict analysis based on segment graph (SG) abstraction, OoO PDES can execute threads safely in parallel and out-of-order (ahead of time) and thus achieves fastest simulation speed but nevertheless maintains the classic SystemC modeling semantics. In order to foster collaboration in the EDA community, we provide the RISC framework as a free open source artifact for full evaluation and possible extension.

For the future, we intend to expand our open source efforts and hope to involve other members of the EDA community to use, evaluate, and extend the RISC framework.

Acknowledgments This work has been supported in part by substantial funding from Intel Corporation for two projects titled "Out-of-Order Parallel Simulation of SystemC Virtual Platforms on Many-Core Architectures" and "Scaling the Recoding Infrastructure for Parallel SystemC Simulation." The authors thank Intel Corporation for the valuable support.

References

1. Accellera Systems Initiative, Core SystemC Language and Examples. http://accellera.org/downloads/standards/systemc
2. Center for Embedded and Cyber-physical Systems, Recoding Infrastructure for SystemC (RISC). http://www.cecs.uci.edu/~doemer/risc.html
3. Center for Embedded and Cyber-physical Systems, RISC Docker Container. https://hub.docker.com/r/ucirvinelecs/risc050/
4. Center for Embedded and Cyber-physical Systems, RISC Release version 0.5.0. http://www.cecs.uci.edu/~doemer/risc.html#RISC050
5. W. Chen, X. Han, R. Dömer, Multi-core simulation of transaction level models using the system-on-chip environment. IEEE Des. Test Comput. **28**(3), 20–31 (2011)
6. W. Chen, X. Han, C.W. Chang, G. Liu, R. Dömer, Out-of-order parallel discrete event simulation for transaction level models. IEEE Trans. Comput. Aided Des. Integr. Circuits Syst. **33**(12), 1859–1872 (2014). https://doi.org/10.1109/TCAD.2014.2356469
7. R. Dömer, Seven obstacles in the way of standard-compliant parallel SystemC simulation. IEEE Embed. Syst. Lett. **8**(4), 81–84 (2016). https://doi.org/10.1109/LES.2016.2617284
8. R. Dömer, G. Liu, T. Schmidt, Parallel simulation, in *Handbook of Hardware/Software Codesign* ed. by S. Ha, J. Teich (Springer, Dordrecht, 2017), pp. 1–32
9. R. Dömer, Z. Cheng, D. Mendoza, A. Dingankar, RISC: recoding infrastructure for SystemC, open source framework for parallel simulation, in *Workshop on Open-Source EDA Technology (WOSET) at ICCAD* (2018)
10. Evaluate Collaboratory, Artifact Evaluation. http://evaluate.inf.usi.ch/artifacts
11. P. Ezudheen, P. Chandran, J. Chandra, B.P. Simon, D. Ravi, Parallelizing SystemC kernel for fast hardware simulation on SMP machines, in *PADS '09: Proceedings of the 2009 ACM/IEEE/SCS 23rd Workshop on Principles of Advanced and Distributed Simulation* (2009), pp. 80–87

12. R. Fujimoto, Parallel discrete event simulation. Commun. ACM **33**(10), 30–53 (1990)
13. IEEE Computer Society, *IEEE Standard 1666-2011 for Standard SystemC Language Reference Manual* (IEEE, New York, 2011)
14. S. Krishnamurthi, Artifact Evaluation Process. http://www.artifact-eval.org/
15. S. Krishnamurthi, J. Vitek, The real software crisis: repeatability as a core value. Commun. ACM **58**(3), 34–36 (2015). https://doi.org/10.1145/2658987
16. G. Liu, T. Schmidt, Z. Cheng, D. Mendoza, R. Dömer, RISC compiler and simulator, release V0.5.0: out-of-order parallel simulatable SystemC subset. Technical Report, CECS-TR-18-03, Center for Embedded and Cyber-physical Systems, University of California, Irvine (2018)
17. D. Mendoza, R. Dömer, A tool for visualization of SystemC models. Technical Report, CECS-TR-17-06, Center for Embedded and Cyber-physical Systems, University of California, Irvine (2017)
18. D.J. Quinlan, ROSE: compiler support for object-oriented frameworks. Parallel Process. Lett. **10**(2/3), 215–226 (2000)
19. C. Roth, S. Reder, H. Bucher, O. Sander, J. Becker, Adaptive algorithm and tool flow for accelerating SystemC on many-core architectures, in *Digital System Design (DSD), 17th Euromicro Conference* (2014)
20. T. Schmidt, G. Liu, R. Dömer, Exploiting thread and data level parallelism for ultimate parallel SystemC simulation, in *Proceedings of the Design Automation Conference (DAC)* (2017)
21. R. Sinha, A. Prakash, H.D. Patel, Parallel simulation of mixed-abstraction SystemC models on GPUs and multicore CPUs, in *Proceedings of the Asia and South Pacific Design Automation Conference (ASPDAC)* (2012)
22. N. Ventroux, T. Sassolas, A new parallel SystemC kernel leveraging manycore architectures, in *Proceedings of the Design, Automation and Test in Europe (DATE) Conference* (2016)
23. J.H. Weinstock, R. Leupers, G. Ascheid, D. Petras, A. Hoffmann, SystemC-link: parallel SystemC simulation using time-decoupled segments, in *Proceedings of the Design, Automation and Test in Europe (DATE) Conference* (2016)
24. D. Yun, J. Kim, S. Kim, S. Ha, Simulation environment configuration for parallel simulation of multicore embedded systems, in *Proceedings of the Design Automation Conference (DAC)* (2011), pp. 345–350

Chapter 8
Impact of Negative Capacitance Field-Effect Transistor (NCFET) on Many-Core Systems

Hussam Amrouch, Martin Rapp, Sami Salamin, and Jörg Henkel

8.1 Introduction

More than a decade ago, the semiconductor technology had entered the so-called nano-CMOS era, in which the transistor's feature sizes became below 90 nm. Since then, the prior trend of voltage scaling came to an end leading to the discontinuation of Dennard's scaling [7]. In Dennard's scaling, both the dimensions of transistor and the operating voltage are typically scaled by the same factor in order to ensure a constant electric field. Due to the non-scalable voltage, ever-increasing power densities in chips became a substantial obstacle for technology scaling due to the limited ability of existing cooling solutions to dissipate the generated heat [8]. To overcome this fundamental problem, the maximum frequency of processors had stopped increasing with every new generation in order to keep the on-chip power densities under acceptable levels and since 2005 the era of many-core processors had started.

To understand the inability of technology to scale voltage, we need to understand what determines the speed of a processor. As a matter of fact, the drive current (ON current) of a transistor dictates its switching speed and hence it ultimately determines the maximum delay of logic paths that form the processor's netlist. The ON current of a transistor is proportional to $(V_{DD} - V_T)$, where V_T denotes the threshold voltage of transistor and V_{DD} denotes the operating voltage. In order to maintain the same level of current, while V_{DD} is scaled down, V_T must also be reduced by almost the same amount. However, reducing V_T comes with an exponential increase in the leakage current (OFF current) of transistor. This is primarily because that the sub-threshold swing of transistor is fundamentally

H. Amrouch · M. Rapp · S. Salamin · J. Henkel (✉)
Karlsruhe Institute of Technology (KIT), Karlsruhe, Germany
e-mail: amrouch@kit.edu; henkel@kit.edu

© The Author(s) 2021
J.-J. Chen (ed.), *A Journey of Embedded and Cyber-Physical Systems*,
https://doi.org/10.1007/978-3-030-47487-4_8

limited to 60 mV/decade at room temperature akin to "Boltzmann tyranny" [21]. Such a fundamental limit inevitably restricts the minimum possible V_T to be at least 300 mV. To ensure a reliable operation, different kinds of safety margins need to be added on top of the minimum voltage, which enforces the operating voltage to remain almost the same with every new technology generation. As above-mentioned, the inability to scale voltage has led to the discontinuation of Dennard's scaling, which, in turn, had led to preventing the frequency of processors from increasing.

In summary, the fundamental limit of sub-threshold swing of transistor is the primary reason behind not scaling voltage and it is the origin of on-chip power density problems that processor's designers are facing since more than a decade ago.

8.1.1 Negative Capacitance Field-Effect Transistor (NCFET)

NCFET integrates a ferroelectric layer inside the gate stack of a transistor, which acts as a negative capacitance. Such a layer provides an amplification of the vertical electric field that the transistor perceives. This, in turn, allows the transistor to overcome the fundamental limitation of sub-threshold swing of 60 mV at room temperature. The principle of NCFET was first proposed in 2008 by S. Salahuddin and S. Datta [16]. After which, it very rapidly gained a large popularity due to the remarkable steep switching and high ON current of transistors [1]. Many experiments have consistently proved NCFET [10]. A breakthrough has recently occurred when GlobalFoundries demonstrated NCFET-based circuits using their state-of-the-art industrial 14 nm FinFET technology [9]. This showed, for the first time, that NCFET technology has become compatible with the existing CMOS fabrication process. In fact, such a compatibility is essential for any emerging technology to be adopted by semiconductor companies. Otherwise, massive production will never be possible.

In practice, NCFET technology enables the transistor to reach the same ON current, without increasing the OFF current, but at a much lower voltage [2]. This is only possible due to steeper sub-threshold swing. Therefore, in an NCFET technology, the processor can still meet the same performance (as in the conventional FET) but at a lower operating voltage leading to a significant power saving. Beside the *low-power* usage scenario of NCFET, *high-performance* usage scenario does also exist. NCFET enables the processor to be clocked at a higher frequency (compared to the conventional FET), while it still be operated at the same voltage due to the increase in the ON current. NCFET technology comes with an important side effect in which it increases the total capacitance of transistor. Such an increase can lead to reliability problems caused by IR-drop and voltage fluctuation during circuit's operation [2, 18]. At the same time, because NCFET technology enables circuits to operate at lower voltages, it is expected that other reliability problems, related to lifetime, to become much less because all the underlying aging mechanisms, such as negative bias temperature instability (BTI) and hot-carrier injection (HCI), strongly depend on the operating voltage [20].

In the following sections, we explain how modeling the NCFET effects from physics all the way to the system level can be done. Then, we explore how a many-core system can profit from the NCFET technology. Finally, we explore the impact that NCFET has on power management schemes and how existing assumptions w.r.t voltage-leakage dependency become not valid anymore when it comes to NCFET, which creates the necessity to develop novel power management techniques.

8.2   Modeling NCFET at the System Level

In the following we provide an overview of how NCFET is modeled at the system level, i.e., for the purpose of simulating many-core processors. Fundamentally, the properties of the ferroelectric layer are modeled at the physics level [12]. Figure 8.1 presents our methodology in which we traverse all layers from physics, through device, gate, and processor level, to model NCFET at the system level. The behavior of transistors with varying thickness of the ferroelectric layer is modeled following the industrial-standard compact model (BSIM-CMG) [5, 14]. Based on this model, we created NCFET-aware cell libraries supporting four different thicknesses of the ferroelectric layer under a wide range of the operating voltage [1]. The thickness ranges from 1 nm (called TFE1) up to 4 nm (TFE4). We then implemented a single many-core tile to the GDSII level and performed timing and power signoffs. The results are explained in detail in the next section. Signoff tools allow to compare power and performance of a processor implemented in different NCFET configurations and are used to extract frequency-dependent scaling factors for dynamic and leakage power. These factors serve as an abstraction at the system level and allow to estimate the power of an NCFET-based processor if the power of

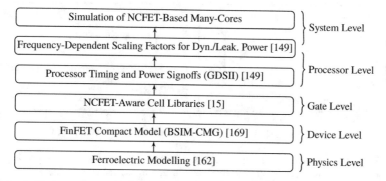

Fig. 8.1 Modeling NCFET at the system level (many-core processors) requires to traverse the whole stack from the physics level, where the effects of the ferroelectric layer are modeled, to the system level, where performance and power of many-core processors are affected

a baseline implementation (conventional FinFET) is known. Finally, these factors are used to simulate a many-core processor (further details in Sect. 8.2.2).

8.2.1 Processor-Level Investigation

This section shows how NCFET affects the performance and power of a single processor. The insights gained from this evaluation are important to build system-level NCFET models and explain observations from system-level simulations. We implemented the layout (GDSII level) of a single tile of the *OpenPiton* many-core [3], which contains a CPU, caches, and a NoC router. Power and timing signoffs are performed for different NCFET configurations (TFE1 to TFE4) and different operating voltages. Further details of the experimental setup can be found in [15].

Figure 8.2a shows how NCFET increases the performance of a processor. It allows to clock a processor at a higher frequency at the same operating voltage or allows to reduce the voltage while still maintaining the same performance (frequency). This is due to the inherent voltage amplification provided by the additional

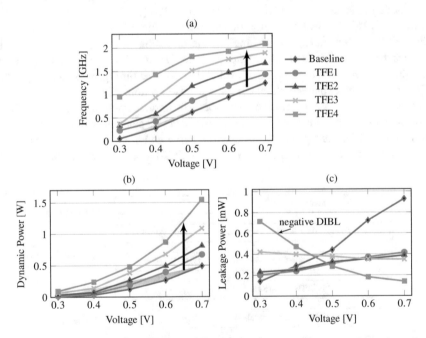

Fig. 8.2 (a) NCFET increases the frequency of a processor at a certain operating voltage, but (b) also increases the dynamic power consumption due to the increase in the transistor gate capacitance and frequency. (c) While leakage increases almost linearly with the operating voltage with conventional FinFET (baseline), this dependency gets weaker with a thin ferroelectric layer and even reverses with TFE4 due to a negative DIBL effect

ferroelectric layer. Like explained earlier, the ferroelectric layer increases the total gate capacity. Together with increased frequency, this increases the dynamic power consumption (Fig. 8.2b). The thicker the ferroelectric layer gets, the higher get the gains in the frequency, but also the higher gets the dynamic power. Figure 8.2c shows that leakage power is affected more severely. NCFET fundamentally changes the trend. With conventional FinFET (baseline), leakage power increases strongly with increasing voltage. When a thin ferroelectric layer is added (TFE1 and TFE2), this dependency becomes weaker, until at TFE3, leakage is almost independent of the voltage. With a thick ferroelectric layer (TFE4), an effect called negative drain-induced barrier lowering (negative DIBL) reverses the leakage dependency on the voltage [13]. Here, leakage increases at lower voltages. We will explain later (Sect. 8.3.3) how this necessitates developing novel power management techniques.

8.2.2   Simulation of NCFET-Based Many-Core

We use the *Sniper* many-core simulator [6] to simulate many-core processors. *McPAT* [11] is used to periodically estimate the power consumption of each core. Since *McPAT* does not support NCFET, it is used to estimate the power with conventional FinFET instead. We develop frequency-dependent scaling factors for dynamic and leakage power based on the processor-level investigation explained earlier.

Figure 8.3 shows the dynamic and leakage power of the single processor studied in the previous section depending on the *frequency*, as opposed to voltage like in Fig. 8.2. Two effects play a role for the dynamic power: NCFET technologies increase the dynamic power at a certain operating voltage (Fig. 8.2b), but also

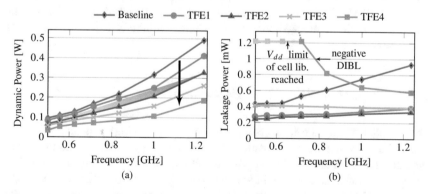

Fig. 8.3 (**a**) While NCFET technologies increase the dynamic power at iso-voltage, they also lower the required operating voltage at iso-frequency, which in total decreases the dynamic power at the same frequency. (**b**) NCFET technologies with a thin ferroelectric layer lower the leakage power, whereas leakage increases with a thick layer (TFE4). Most importantly, the negative DIBL effect reverses the leakage dependency, where lowering the V/f-levels increases leakage

allow to go to a lower operating voltage (Fig. 8.2a) while still maintaining the same frequency. Lowering the operating voltage has a stronger effect on the dynamic power. Consequently, NCFET technologies lower the dynamic power when operating at the same frequency (Fig. 8.3a). Figure 8.3b shows how leakage power depends on the V/f-level. The reverse leakage dependency with TFE4 strongly increases the leakage power. Below 700 MHz, TFE4 would allow to reduce the voltage below 0.2 V, which is the lower limit of the cell library.

Figure 8.3a,b allows to estimate the dynamic and leakage power consumption of a processor that is implemented in NCFET, if the power consumption in the baseline (conventional FinFET) is known. We extract frequency-dependent scaling factors for both dynamic and leakage power. These factors serve as an abstraction that allows simulation of complex benchmark applications, like *PARSEC* [4], on many-core processors with dozens of cores. We thereby scale the leakage and dynamic power that is estimated by *McPAT* to estimate the power consumption of NCFET-based many-cores. For brevity, details on this approach are omitted here and can be found in [15].

8.3 Performance, Power, and Cooling Trade-Offs with NCFET-based Many-Cores

NCFET fundamentally changes the characteristics of transistors and therefore also changes the performance and power of circuits [19], single-core processors [1], and many-core processors [15]. This section demonstrates the impact of the thickness of the ferroelectric layer on the power and performance of a many-core processor. We show that the optimal thickness depends on many factors, such as the application characteristics and the cooling scenario. This section evaluates performance, power, and cooling of a 25-core many-core operating under a thermal constraint of 80°C. We study *PARSEC* [4] tasks with up to eight slave threads. Their characteristics range from highly memory-bound (e.g., *canneal*) to highly compute-bound (e.g., *swaptions*).

8.3.1 Impact of NCFET on Performance

Due to high power densities (failure of Dennard's scaling) and limited cooling capabilities, it is not always possible in modern technology nodes to simultaneously operate all cores at the peak V/f-levels without violating the thermal constraint. This study investigates the use-case in which cores with an active thread are operated at the peak V/f-levels and cores without a thread mapped to it are power-gated. In this use-case, four factors affect the thermally sustainable utilization (i.e., the number of cores that can be turned on): the application characteristics (power consumption), the mapping of threads to cores, the cooling system, and the transistor technology.

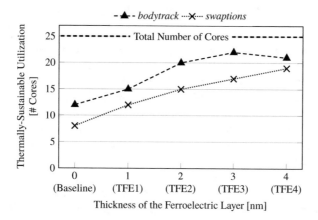

Fig. 8.4 NCFET technologies increase the thermally sustainable utilization of a 25-core many-core, i.e., the number of usable cores without violating the thermal constraint, compared to the baseline (conventional FinFET). The optimal thickness of the ferroelectric layer depends on the application characteristics

We use an Integer Linear Program to obtain the thermally-optimal mapping of threads to cores, which minimizes the formation of hotspots and, thereby, maximizes the thermally sustainable utilization. We study the use-case of a passive cooling, i.e., there is no fan on top of the heat sink.

Figure 8.4 shows the thermally sustainable utilization of two benchmarks *bodytrack* and *swaptions* during the parallel section of the benchmarks (Region of Interest) for different NCFET technologies. Other benchmarks are available in [15]. *Swaptions* is a highly compute-intensive task, which results in high power consumption and therefore, the thermally sustainable utilization in the baseline is low (only 8 out of 25 cores). Dynamic power forms the major part of the total power consumption and therefore, thicker ferroelectric layers increase the thermally sustainable utilization because dynamic power is reduced (compare Fig. 8.3a). Consequently, the highest performance is observed with the thickest ferroelectric layer (TFE4). *Bodytrack* is less compute-intense and has lower dynamic power consumption and consequently lower total power. This results in a higher thermally sustainable utilization compared to *swaptions*. However, due to lower dynamic power, leakage power accounts for a larger fraction of the total power. As demonstrated in Fig. 8.3b, TFE4 increases the leakage significantly over TFE3. Consequently, TFE4 results in a lower thermally sustainable utilization than TFE3 for *bodytrack* and the highest performance is observed with TFE3.

These investigations show that *the optimal thickness of the ferroelectric layer depends on the application characteristics*. Further investigations on how NCFET affects the performance in the case that cores are not operated at the peak V/f-levels can be read in [15]. These investigations additionally study forced-convection cooling (a heat sink with a fan) and reveal that *the optimal thickness of the ferroelectric layer also depends on the cooling scenario*.

8.3.2   Impact of NCFET on Cooling Requirements

This section studies how NCFET reshapes the existing trade-off between cooling costs and achievable performance, where higher performance comes at the cost of higher power dissipation and therefore higher cooling costs. We study the use-case in which the many-core is operated at its peak performance, i.e., all cores are active at the peak V/f-levels and determine the cooling capabilities that are required to make this use-case thermally safe. The cooling capabilities are measured by the inverse of thermal resistance of the heat sink $1/R_{th}$. Varying this value corresponds to changing the air convection.

Figure 8.5 shows the required cooling capabilities for the three *PARSEC* benchmarks *swaptions*, *bodytrack*, and *canneal* during the parallel section of the benchmarks (Region of Interest). NCFET technologies allow to reduce the cooling capabilities over the baseline (conventional FinFET). Most importantly, the required cooling capabilities are minimized at different thicknesses of the ferroelectric layer depending on the application. *Swaptions* is highly compute-intensive and consequently, dynamic power accounts for the majority of the total power. Increasing the thickness of the ferroelectric layer reduces the dynamic power (see Fig. 8.3) and therefore reduces the required cooling. *Canneal* on the other side is highly memory-bound and therefore, the power consumption is dominated by leakage. Leakage is minimized at TFE2, which consequently minimizes the cooling requirements. *Bodytrack* shows intermediate values for the dynamic power and therefore, TFE3 is optimal. *This investigation shows again that the optimal thickness of the ferroelectric layer depends on the application characteristics and ranges from 2 nm to 4 nm.*

Fig. 8.5 NCFET technologies decrease the required cooling capabilities while maintaining the same maximum temperature of 80°C under full system utilization (all cores active at peak V/f-levels). The thickness of the ferroelectric layer that results in the lowest cooling costs depends on the application characteristics and ranges from 2 nm (with *canneal*) to 4 nm (with *swaptions*)

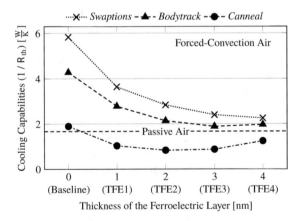

8.3.3 Impact of NCFET on Power Management Techniques

The above investigations use the well-established concept of V/f-pairs that are determined at design time by selecting the operating voltage for a given frequency as the lowest voltage that makes operating at this frequency reliable. This is a reasonable approach with conventional transistor technologies, because using a higher voltage would unnecessarily increase both dynamic and leakage power. However, this is no longer true with NCFET with a thick ferroelectric layer (TFE4). Here, increasing the voltage decreases the leakage power. This leads to new optimization potential by selecting the operating voltage for a given frequency, which is demonstrated in the next section.

8.4 NCFET-Aware Voltage Scaling

Dynamic voltage scaling (DVS) technique for processor power management is considered to be one of the most effective ways to reduce the energy consumption of an application. DVS technique typically selects the minimum operating voltage V_{min} that sustains the operating frequency of the processor at runtime based on the frequency demands of the application being executed. Reducing the operating voltage, in conventional FET, results in reducing the total power consumption, which implicitly reduces both dynamic and leakage power. However, such a well-known voltage dependency becomes inverse with respect to leakage power in NCFET due to the negative DIBL effect (see Sect. 8.2.1). With such opposed dependencies (dynamic and leakage) to the operating voltage, total power follows the dominant component when voltage changed which leads to a novel trade-off. Consequently, power is not necessarily minimized at the minimum voltage V_{min}, which traditional DVS selects, but at another voltage V_{opt}. Unawareness of NCFET and its trade-off could lead to not minimize the total power consumption. Therefore, in this section, a novel NCFET-aware voltage scaling technique is presented [17] to overcome the shortness that traditional DVS has in NCFET-based processors.

8.4.1 Importance of NCFET-Aware DVS

With traditional DVS, a set of voltage-frequency pairs are typically selected at design time and later are employed by the DVS technique at runtime to optimize the power. In this case, the lower the selected voltage is, the lower the total power is. Due to the new inverse dependency in leakage power that NCFET exhibits, this is not always valid with respect to NCFET. To demonstrate the consequence of such an inverse dependency at the system level, we plot the total power consumption and its components of the master thread of PARSEC *canneal* benchmark in Fig. 8.6.

Fig. 8.6 Total power and its components (i.e., leakage and dynamic) of *canneal* master thread starting from the minimum voltage V_{min} required to sustain 1.0 GHz frequency. The total power is not minimized at V_{min}. The operating voltage required to minimize total power V_{opt} appears at a higher voltage than V_{min} due to leakage increases in NCFET

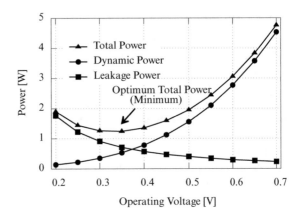

The power examined starting from V_{min}, that traditional DVS selects to sustain the required frequency, and then to overscale the operating voltage. The result shows that the power is not minimized at V_{min} (i.e., $V_{opt} \neq V_{min}$).

Different workloads exhibit different characteristics and hence different total power. Therefore, the contribution of power components differs. Traditional DVS neglects this difference as both contributions (leakage and dynamic) are affected in the same manner with voltage (both are reduced). With NCFET, the contribution of leakage to the total power cannot be neglected because it affects the operating voltage selection when DVS tries to minimize total power. Hence, based on the leakage share, V_{opt} could differ from V_{min}.

For the aforementioned reasons, NCFET-aware DVS is crucial due to the change in the behavior of total power consumption over voltage scaling which emerges from the inverse dependency with respect to leakage power in NCFET.

8.4.2 NCFET-Aware DVS Technique

To enable runtime voltage selection, DVS first needs to determine workload characteristics and then V_{opt} can be correctly selected. Therefore, determining V/f-pairs at runtime, like in traditional DVS techniques, is not possible here. Instead, the results from Sect. 8.2.1 have been used to build the power and performance analytical models at design time. Then, these models can be integrated with our new NCFET-aware DVS technique for runtime voltage selection.

8.4.2.1 Design-Time Models

Power and Performance Modeling The maximum operating frequency $f_{max}(V)$ depends on the voltage V over the minimum delay $d_{min}(V)$:

$$d_{\min}(V) = a_{del} \cdot V^{b_{del}} + c_{del}; \quad f_{\max}(V) = \frac{1}{d_{\min}(V)} \tag{8.1}$$

$a_{del} > 0$, $b_{del} < 0$, $c_{del} \geq 0$ are constants fitting parameters obtained at design time. Peak leakage and peak dynamic power consumption results by operating at maximum frequency are

$$P_{leak}(V) = a_{leak} \cdot V^{b_{leak}} \tag{8.2}$$

$$P_{dyn}^{peak}(V, d_{\min}(V)) = a_{dyn} \cdot V^{b_{dyn}} + c_{dyn} \tag{8.3}$$

$a_{dyn} > 0$, $b_{dyn} > 1$, $c_{dyn} \geq 0$, $a_{leak} > 0$, $b_{leak} < 0$ are constant fitting parameters obtained at design time. Both $P_{dyn}^{peak}(V, d_{\min}(V))$ and $P_{leak}(V)$ are *convex* in V. By lowering the operating frequency of the CPU (higher delay), dynamic power decreases. However, since leakage power is independent from CPU activity, it is not affected.

$$P_{dyn}^{peak}(V, d) = \frac{d_{\min}(V)}{d} \cdot P_{dyn}^{peak}(V, d_{\min}(V)) \tag{8.4}$$

Therefore, $P_{dyn}^{peak}(V, d)$ is convex in V (for constant d) if $b_{dyn} + b_{del} > 1$.

8.4.2.2   Runtime Models

Workload-Dependent Power Modeling Dynamic power consumption $P_{dyn}(V, d)$ is affected by the running workload, which is reduced by a factor $0 \leq r_{dyn} \leq 1$ from the peak dynamic power $P_{dyn}^{peak}(V, d)$:

$$P_{dyn}(V, d) = r_{dyn} \cdot P_{dyn}^{peak}(V, d) \tag{8.5}$$

$$P_{total}(V, d) = P_{dyn}(V, d) + P_{leak}(V) \tag{8.6}$$

r_{dyn} is not constant since it represents the current workload activity. Therefore, total power consumption $P_{total}(V_c, d)$ at the current voltage V_c, r_{dyn} is

$$r_{dyn} = \frac{P_{dyn}(V_c, d)}{P_{dyn}^{peak}(V_c, d)} = \frac{P_{total}(V_c, d) - P_{leak}(V_c)}{P_{dyn}^{peak}(V_c, d)} \tag{8.7}$$

Optimal Voltage Computing V_{opt} that minimizes the total power can be obtained from the power and performance models:

$$V_{\min}(d) = \left(\frac{d - c_{del}}{a_{del}} \right)^{\frac{1}{b_{del}}} \tag{8.8}$$

Algorithm 1 NCFET-aware voltage scaling algorithm to select the optimal voltage (V_{opt}) at runtime [17]

Require: Power and performance models: $P_{dyn}^{peak}(c, d)$ and $P_{leak}(V)$, current operating voltage V_c and delay d, current power consumption P_{curr}, min. voltage resolution ϵ
Ensure: Optimal operating voltage V_{opt}
1: $r_{dyn} \leftarrow (P_{curr} - P_{leak}(V_c)) / P_{dyn}^{peak}(V_c, d)$ ▷ Eq. (8.7)
2: $V_{opt} \leftarrow V_{min}(d)$ ▷ Eq. (8.8)
3: **repeat**
4:     $\Delta V_{opt} \leftarrow -P_{total}(V_{opt}, d)' / P_{total}(V_{opt}, d)''$
5:     $V_{opt} \leftarrow V_{opt} + \Delta V_{opt}$ ▷ iterative update
6:     **if** $V_{opt} < V_{min}(d)$ **then return** $V_{min}(d)$ ▷ out of bounds
7:     **if** $V_{opt} > V_{max}$ **then return** V_{max} ▷ out of bounds
8: **until** $\Delta V_{opt} < \epsilon$ ▷ Termination criteria
9: **return** V_{opt}

$$V_{opt}(d, r_{dyn}) = \underset{V_{min}(d) \leq V \leq V_{max}}{\arg\min} \quad P_{total}(V, d) \tag{8.9}$$

Since $P_{total}(V, d)$ is composed of convex functions, our implemented algorithm exploits that $P_{total}(V, d)$ is convex in V. This guarantees that $P_{total}(V, d)$ has exactly one minimum w.r.t. V within the range $[V_{min}(d), V_{max}]$. Algorithm 1 summarizes our implemented DVS technique and obtaining V_{opt}.

8.4.3 Operating Voltage Selection

Both DVS techniques differ in the way they select the operating voltage. Therefore, to show the different behavior between both techniques in operating voltage selection, the design space of the operating voltage selection with NCFET-aware (V_{opt}) and NCFET-unaware DVS (V_{min}) has been explored in Fig. 8.7. NCFET-unaware DVS sets V_{min} that is needed to sustain the required frequency and therefore workload characteristic is not considered. Contrarily, NCFET-aware DVS considers the workload characteristic as it depends on the ratio of leakage to total power measured at V_{min}. The explored design space in Fig. 8.7 shows two distinct regions: (1) For low leakage to total power ratio and for high frequencies, the same voltage is selected (similar action) by both techniques (i.e., $V_{opt} = V_{min}$). (2) For high ratios of leakage to total power or low frequencies, NCFET-aware DVS selects a higher voltage ($V_{opt} > V_{min}$). Moreover, Fig. 8.7 reveals that: the higher the required frequency or the higher the leakage to total power ratio, the higher V_{opt} is.

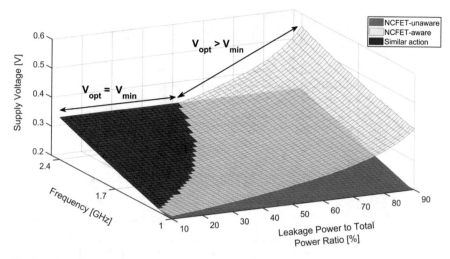

Fig. 8.7 Operating voltage selection using both DVS techniques. Two regions appear: (1) NCFET-aware selection differs from NCFET-unaware ($V_{min} \neq V_{opt}$). (2) Similar action is done by both DVS as they select the same operating voltage ($V_{min} = V_{opt}$). NCFET-unaware DVS selects V_{min} (that sustains the required frequency) and NCFET-aware selects V_{opt} to minimize the power depending on the frequency and the ratio of leakage to total power. NCFET-aware DVS selects higher voltages when leakage power becomes prominent or at lower frequency

8.4.4   Evaluation

8.4.4.1   Experimental Setup

Using the same setup in Sect. 8.2.1, power and delay results were examined using the highest ferroelectric thickness (4 nm). Afterwards, the power and performance analytical models have been developed as described in Sect. 8.4.2.

For system-level simulation, relying on the setup described in Sect. 8.2.2, the NCFET-aware DVS technique (Algorithm 1) has been used to select the operating voltage when a set of tasks were examined from the PARSEC benchmark suite [4]. The frequencies are set between 1.0 GHz and 2.4 GHz. V_{dd} is set between 0.2 V and 0.7 V. The low operating voltages V_{dd} in NCFET are lower than traditional FET due to the inherent voltage amplification in NCFET provided by the negative capacitance. For fair comparisons, simulators for both DVS cases were configured to have: the same frequencies, and architecture, in addition to running the same benchmarks. Hence, only voltage selection differs based on DVS decision.

8.4.4.2   NCFET-Aware DVS Results and Analysis

To show the effectiveness of the NCFET-aware DVS, we first show how NCFET-aware DVS actually operates to save power and later to report the energy savings

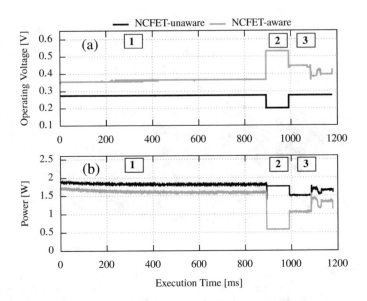

Fig. 8.8 (a) Operating voltage and (b) total power consumption during an interval of the execution time of the *canneal* master thread with NCFET-unaware and NCFET-aware DVS. NCFET-aware DVS selects higher voltage most of the time (in this particular example) and reduces the power further at the same CPU frequency. Voltage selection is based on workload characteristics

for different benchmarks in comparison with NCFET-unaware DVS. Accordingly, an illustrative example of the master thread of PARSEC *canneal* benchmark was selected. Figure 8.8 shows distinct phases during an interval of the execution time. In phase-1, in Fig. 8.8b, it shows the total power consumption when the frequency is set at 1.7 GHz. Traditional DVS sets V_{dd} to the minimum voltage (0.28 V) which required to sustain this frequency. Thus, dynamic power is minimized but the leakage power is not. NCFET-aware DVS sets V_{dd} to a higher value to guarantee a better trade-off. This will increase the dynamic power but strongly decreases leakage power resulting in a power saving. In phase-2, the master thread is idle and waits for the termination of the slave threads. Therefore, frequency is reduced to the minimum frequency (1.0 GHz). Traditional DVS reduces V_{dd} to 0.2 V due to the low required frequency in which it increases the leakage power. NCFET-aware DVS, instead of reducing V_{dd}, increases the voltage to 0.53 V, which decreases the leakage power. Thereby, the total power consumption in phase-2 is reduced by 67 % compared to the traditional DVS. In phase-3, after the slaves terminated, the master resumes operation and its frequency is boosted again to 1.7 GHz. It is worth to mention that the performance obtained with both DVS techniques is the same. This is because they do not affect the frequency, but only set the V_{dd} under performance constraint.

To reveal the energy savings, different PARSEC benchmarks were examined when active threads are operated at 1.7 GHz and idle cores are suppressed to 1.0 GHz. Figure 8.9 summarizes the energy savings. Energy savings range as shown in Fig. 8.9 from 14 % up to 27 % and in average are up to 20 %.

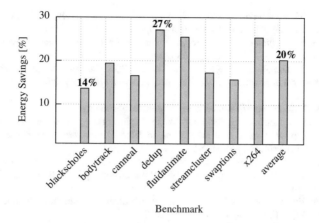

Fig. 8.9 Energy saving results of different benchmarks using the NCFET-aware DVS compared to NCFET-unaware DVS. Energy savings range from 14 % up to 27 % , and in average 20%

8.5   Conclusion

In this chapter, we investigated how NCFET technology impacts the existing trade-offs in processors and how it can reshape the future of many-core systems. Compared to the existing FinFET technology, NCFET technology allows the processor to operate at a much lower voltage while it still meets the same performance. This results in a considerable power saving and as a result the total number of cores, that can be simultaneously turned on at the maximum frequency, increases without violating the predetermined thermal constraints. We also showed how NCFET inverses the leakage-voltage dependency and proposed a new NCFET-aware DVS technique that provides an energy saving of 20% on average compared to conventional DVS techniques, which are unaware of the new leakage-voltage dependency that NCFET brings.

Acknowledgments The authors would like to thank Yogesh S. Chauhan, Girish Pahwa, and Amol Gaidhane from the Indian Institute of Technology Kanpur (IIT Kanpur) for their valuable contribution and support in the NCFET analysis at the device level and compact modeling.

References

1. H. Amrouch, G. Pahwa, A.D. Gaidhane, J. Henkel, Y.S. Chauhan, Negative capacitance transistor to address the fundamental limitations in technology scaling: processor performance. IEEE Access **6**, 52754–52765 (2018)
2. H. Amrouch, S. Salamin, G. Pahwa, A.D. Gaidhane, J. Henkel, Y.S. Chauhan, Unveiling the impact of IR-drop on performance gain in NCFET-based processors. IEEE Trans. Electron Devices **66**(7), 3215–3223 (2019). https://doi.org/10.1109/TED.2019.2916494

3. J. Balkind, M. McKeown, Y. Fu, T. Nguyen, Y. Zhou, A. Lavrov, M. Shahrad, A. Fuchs, S. Payne, X. Liang, M. Matl, D. Wentzlaff, OpenPiton: an open source Manycore research framework, in *Architectural Support for Programming Languages and Operating Systems (ASPLOS)* (2016), pp. 217–232. https://doi.org/10.1145/2872362.2872414
4. C. Bienia, S. Kumar, J.P. Singh, K. Li, The PARSEC benchmark suite: characterization and architectural implications, in *Parallel Architectures and Compilation Techniques (PACT)* (2008), pp. 72–81
5. BSIM-CMG Model. http://bsim.berkeley.edu/models/bsimcmg
6. T.E. Carlson, W. Heirman, L. Eeckhout, Sniper: exploring the level of abstraction for scalable and accurate parallel multi-core simulation, in *High Performance Computing, Networking, Storage and Analysis (SC)* (ACM, New York, 2011), p. 52
7. R.H. Dennard, F.H. Gaensslen, V.L. Rideout, E. Bassous, A.R. LeBlanc, Design of ion-implanted MOSFET's with very small physical dimensions. IEEE J. Solid State Circuits **9**(5), 256–268 (1974)
8. L.B. Kish, End of Moore's law: thermal (noise) death of integration in micro and nano electronics. Phys. Lett. A **305**(3), 144–149 (2002)
9. Z. Krivokapic, U. Rana1, R. Galatage, A. Razavieh, A. Aziz, J. Liu, J. Shi, H. Kim, R. Sporer, C. Serrao, A. Busquet, P. Polakowski, J. Müller, W. Kleemeier, A. Jacob1, D. Brown, A. Knorr, R. Carter, S. Banna, 14 nm ferroelectric FinFET technology with steep subthreshold slope for ultra low power applications, in *IEEE International Electron Devices Meeting (IEDM)* (2017), pp. 15.1.1–15.1.4
10. D. Kwon, K. Chatterjee, A.J. Tan, A.K. Yadav, H. Zhou, A.B. Sachid, R.D. Reis, C. Hu, S. Salahuddin, Improved subthreshold swing and short channel effect in FDSOI n-channel negative capacitance field effect transistors. IEEE Electron Device Lett. **39**(2), 300–303 (2018). https://doi.org/10.1109/LED.2017.2787063
11. S. Li, J.H. Ahn, R.D. Strong, J.B. Brockman, D.M. Tullsen, N.P. Jouppi, The McPAT framework for multicore and manycore architectures: simultaneously modeling power, area, and timing. Trans Archit. Code Optim. (TACO) **10**(1), 5 (2013)
12. J. Müller, T.S. Boscke, U. Schröder, S. Mueller, D. Bräuhaus, U. Böttger, L. Frey, T. Mikolajick, Ferroelectricity in simple binary ZrO_2 and HfO_2. Nano Lett. **12**(8), 4318–4323 (2012). https://doi.org/10.1021/nl302049k
13. G. Pahwa, T. Dutta, A. Agarwal, Y.S. Chauhan, Designing energy efficient and hysteresis free negative capacitance FinFET with negative DIBL and 3.5 XI ON using compact modeling approach, in *European Solid-State Circuits Conference (ESSCIRC)* (2016), pp. 49–54
14. G. Pahwa, T. Dutta, A. Agarwal, S. Khandelwal, S. Salahuddin, C. Hu, Y.S. Chauhan, Analysis and compact modeling of negative capacitance transistor with high ON-current and negative output differential resistance—Part II: model validation. IEEE Trans. Electron Devices **63**(12), 4986–4992 (2016)
15. M. Rapp, S. Salamin, H. Amrouch, G. Pahwa, Y. S. Chauhan, J. Henkel: Performance, power and cooling trade-offs with NCFET-based many-cores, in *Design Automation Conference (DAC)* (2019)
16. S. Salahuddin, S. Datta, Use of negative capacitance to provide voltage amplification for low power nanoscale devices. Nano Lett. **8**(2), 405–410 (2008). https://doi.org/10.1021/nl071804g
17. S. Salamin, M. Rapp, H. Amrouch, G. Pahwa, Y. S. Chauhan, J. Henkel, NCFET-Aware voltage scaling, in *The International Symposium on Low Power Electronics and Design (ISLPED)* (2019)
18. S. Salamin, V.M. van Santen, H. Amrouch, N. Parihar, S. Mahapatra, J. Henkel, Modeling the interdependences between voltage fluctuation and BTI aging. IEEE Trans. Very Large Scale Integr. VLSI Syst. **27**(7), 1652–1665 (2019)
19. S.K. Samal, S. Khandelwal, A.I. Khan, S. Salahuddin, C. Hu, S.K. Lim, Full chip power benefits with negative capacitance FETs, in *International Symposium Low Power Electronics and Design (ISLPED)* (2017)

20. V.M. van Santen, H. Amrouch, J. Henkel, Modeling and mitigating time-dependent variability from the physical level to the circuit level. IEEE Trans. Circuits Syst. I Regul. Pap. **66**(7), 2671–2684 (2019)
21. V.V. Zhirnov, R.K. Cavin, Nanoelectronics: negative capacitance to the rescue? Nat. Nanotechnol. **3**(2), 77–78 (2008)

Chapter 9
Run-Time Enforcement of Non-functional Program Properties on MPSoCs

Jürgen Teich, Pouya Mahmoody, Behnaz Pourmohseni, Sascha Roloff, Wolfgang Schröder-Preikschat, and Stefan Wildermann

9.1 Introduction

In a broad range of embedded systems, e.g., in real-time and safety-critical domains, applications require guarantees (rather than a best-effort behavior) w.r.t. non-functional properties of their execution such as timing characteristics and reliability. Delivering the required guarantees is, therefore, of utmost importance for the successful introduction of multi-/many-core architectures in the embedded domains of computing. In a many-core context, existing analysis tools either impose an immense computational complexity or deliver worst-case guarantees that suffer from a massive over-/under-approximation for the vast majority of execution scenarios (due to the inherent uncertainty of these scenarios) and, hence, are of no practical interest. Noteworthy, a major source of this uncertainty originates from the interferences among concurrent applications.

In view of abundant computational and storage resources becoming available, new programming paradigms such as *invasive computing* [22] have proved effective in alleviating these interferences by means of spatial isolation among applications. Here, hybrid (static analysis/dynamic mapping) approaches, e.g., [11, 19, 20, 24], enable a static generation of different mappings for each application on system resources in form of mapping classes rather than individual mappings. For each concrete mapping within such a class, safe bounds on the non-functional execution properties, e.g., latency, may hold, see, e.g., [25]. The statically generated and analyzed sets of optimal mapping classes are then provided to the run-time system which checks the availability of such constellations of resources under the current system workload, and, if enough resources are available, finally launches the

J. Teich (✉) · P. Mahmoody · B. Pourmohseni · S. Roloff · W. Schröder-Preikschat · S. Wildermann
Friedrich-Alexander-Universität Erlangen-Nürnberg, Erlangen, Germany
e-mail: juergen.teich@fau.de

© The Author(s) 2021
J.-J. Chen (ed.), *A Journey of Embedded and Cyber-Physical Systems*,
https://doi.org/10.1007/978-3-030-47487-4_9

application [25]. Such a hybrid approach has been implemented within the language InvadeX10, a library-based extension of the X10 programming language. In this extension, the so-called *requirements* [23] on non-functional execution properties, e.g., latency, may be annotated to individual applications or program segments thereof.

Although spatial isolation among applications significantly reduces the afore-mentioned uncertainties, a considerable degree of them remain unaffected which might be unacceptable, e.g., for safety-critical applications. But also, real-world applications from the domain of streaming often exhibit a large jitter in the latency and throughput (in spite of inter-application resource isolation) which is not tolerable, e.g., in case of camera-based medical surgery. This intolerable or annoying variation mainly stems from two sources of uncertainty that cannot be eliminated or restricted through resource isolation:

- **Execution State Uncertainty.** This source of uncertainty originates either from the environment (termed exogenous), e.g., ambient temperature, or from within the computing system itself (termed endogenous), e.g., cache states or the voltage/frequency modifications applied by the power manager. While the vast majority of exogenous sources of uncertainty cannot be avoided or controlled, endogenous sources of uncertainty may be eliminated, e.g., by flushing caches before execution or by pinning the voltage/frequency of each core to a desired fixed level.
- **Input Uncertainty.** This source of uncertainty originates from the application's input(s). For instance, in image processing, the content of a scene may greatly influence the amount of workload to be processed per image.

In the presence of execution state and input uncertainties, *application-specific run-time techniques* can offer a practical approach to confine the non-functional properties of execution within acceptable bounds or to prevent the violation of requirements. Such techniques dynamically adjust a given set of control knobs, e.g., voltage/frequency settings, in reaction to observed (or predicted) changes in the input and/or environment states to steer the non-functional properties of execution within the desired range. Examples of such approaches include the enforcement of safety properties using automata [13] or the satisfaction of timing constraints (while minimizing energy) using control-theory oriented approaches [9]. We refer to this emerging class of application-specific run-time techniques as *Run-Time Requirement Enforcement (RRE)*. This paper presents the fundamentals, definitions, and taxonomy of RRE in the context of many-core systems. We exemplify the practice of different classes of RRE techniques and present a discussion on their advantages, drawbacks, and challenges in a case study on the enforcement of timing requirements for a distributed real-time image processing application.

9.2   Preliminaries and Definitions

9.2.1   System Model

A many-core *architecture* is typically organized as a set of so-called storage, I/O, and compute tiles which are interconnected by a Network-on-Chip (NoC) for scalability, see, e.g., Fig. 9.1. Memory and I/O tiles enable mass storage and off-chip access, respectively. Each compute tile is typically organized as a multi-core or a processor array and comprises a set of processing cores, peripherals such as memories, and a network adapter which are interconnected via one or more buses. An *application* to be executed on the architecture is typically composed of a set of processing tasks with known data dependencies, provided as a task graph. In case of periodic applications, actor-based models of computation and languages such as ActorX10 [17] may be used for parallel programming of MPSoCs. Each application may be augmented with one or a set of *requirements* on specific non-functional properties of its execution, e.g., execution time, throughput, or power corridors. In the following, a *mapping* of an application on a given architecture corresponds to a binding of its tasks to platform cores, a routing of the data exchanged between communicating tasks, an allocation of the required processing, communication, and storage resources, and a scheduling of tasks and communications on the allocated

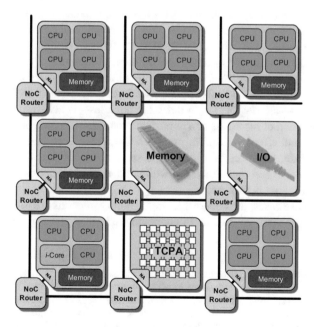

Fig. 9.1   A heterogeneous invasive MPSoC architecture

resources. Alternatively to concrete mappings, a set of constraints that reflect a constellation of required resources and, hence, correspond to several concrete deployments of the application on the architecture may be characterized at design time through techniques of design space exploration [19, 24, 25].

9.2.2 *-Predictability

Non-functional requirements of applications, e.g., real-time constraints, can often be expressed in form of intervals according to the definition for the predictability of a non-functional property from [23]:

Definition 9.1 (*-predictability) Let o denote a non-functional property of a program (implementation) p and the uncertainty of its input (space) given by I and environment by Q. The predictability (marker) of objective o for program p is defined by the interval

$$o(p, Q, I) = [inf_o(p, Q, I), \ldots, sup_o(p, Q, I)] \qquad (9.1)$$

where inf_o and sup_o denote the infimum and supremum of property o, respectively, under variation of state $q \in Q$ and input $i \in I$.

Figure 9.2 exemplifies Definition 9.1 for three implementations p_1, p_2, and p_3 of an application with two requirements in terms of latency and power consumption.[1] The rectangle associated with each implementation p_i confines the observable latency and power range for p_i under the variation of input $i \in I$ and state $q \in Q$. As illustrated, p_1 never satisfies the latency requirement under any input/state and, thus, is of no interest. Contrarily, p_2 satisfies both requirements in all input/state scenarios which—although offering desirable qualities—is achieved through, e.g., an over-reservation of resources or a persistently maximized core voltage/frequency which is often not affordable and/or practical. Contrarily to p_1 and p_2, p_3 exhibits an attractive case: Under certain input/state scenarios it satisfies the requirements (with an affordable resource demand), while under other scenarios the acceptable latency-power region is surpassed.

In real-life use cases, the observable predictability intervals are often too coarse, so that a large share of viable implementations (like p_3) do not satisfy the

---

[1]Note that a lower bound on latency makes sense in many applications that communicate result data to other applications or systems. Here, either buffer limitations would cause overflows in case the producer would be faster than the consumer. Alternatively, data might get lost if the producer overwrites not yet consumed data. Similarly, a minimal lower bound is the default in the case of reliability requirements. There, the lower bound could indicate a minimal expected lifetime. Finally, even lower power bounds can be found in the area of high-performance computing. In fact, the energy bill of a supercomputer increases by the amount of not consumed power but reserved by the provider.

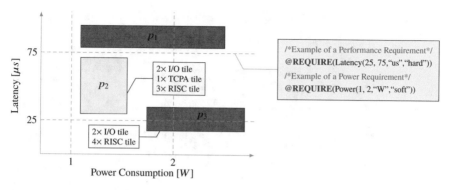

Fig. 9.2 Example of a program p with a latency requirement and a power requirement given each by an interval (corridor). Shown are three program implementations. p_1 does not satisfy the latency requirement for any possible execution. p_2 satisfies the two requirements for any possible variation in input $i \in I$ and state $q \in Q$. Finally, p_3 may satisfy the two requirements, but obviously not for all observable executions. Here, run-time requirement enforcement techniques might be applicable to control the resources of the platform based on run-time monitoring to stay within the requirement corridors

given requirements under all input/state scenarios. For such partially satisfactory implementations, run-time techniques can be employed to render them consistently satisfactory by regularly monitoring (or predicting) the online input/state scenario and either acting proactively to avoid any violation of a set of given requirements, e.g., by adjusting the voltage/frequency settings of cores prior to program execution, or in reaction to any observed violation. The purpose of such run-time techniques is, therefore, to enforce that the desired latency and power corridor are never (or only occasionally) violated. We refer to these application-specific run-time techniques as Run-Time Requirement Enforcement (RRE) in the following.

9.3   Run-Time Requirement Enforcement

To satisfy a set of given requirements, the observable predictability intervals of the partially satisfactory implementations must be obviously reduced. In general, this can be achieved by techniques such as *restricting* the input space I or using *approximate computing* [23]. Alternatively, *isolation* techniques that reduce the state space Q may be applied such as the use of simpler cores, resource reservation protocols, or using *invasive computing* [22]. In the latter approach, an application program invades a set of processing and communication resources prior to execution. Through inter-application isolation, composability is established which is essential for an independent analysis of individual applications [1, 8, 12].

Definition 9.2 (Run-Time Requirement Enforcer (RRE)) A Run-Time Requirement Enforcer (RRE) of a requirement $r_o(p) = [LB_o, UB_o]$ of a program p is a

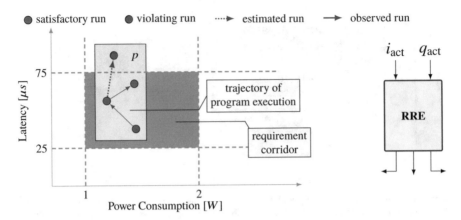

Fig. 9.3 Example of Run-Time Requirement Enforcement (RRE)

control technique to steer o within the corridor spanned by a lower bound LB_o and an upper bound UB_o for each execution of p.

Figure 9.3 exemplifies Definition 9.2 for a latency and a power requirement corridor of an implementation p of an application. An RRE is also depicted whose task is to confine the observable predictability interval of p within the corridor specified by the latency and power requirements. Given the actual (current) input $i_{act} \in I$ and state $q_{act} \in Q$, the RRE in this case proactively estimates the expected latency L_{est} and power consumption P_{est} based on which it takes actions (outgoing arcs of the RRE) with the goal to avoid any violation of the requirements. Examples of RRE actions include adjusting the voltage/frequency of the cores or awaking reserved cores that are currently in a sleep state for power reduction, or even changing the mapping of some tasks to other cores [14].

9.4 Taxonomy of Run-Time Requirement Enforcers

According to [23], each requirement of an application can be either *soft* or *hard*. In case of a soft requirement, occasional violations are still considered acceptable. In this context, a RRE can be classified as either a *loose* or a *strict* enforcement technique as follows:

Definition 9.3 (Loose/Strict RRE) A Run-Time Requirement Enforcer (RRE) of a requirement $r_o(p) = [LB_o, UB_o]$ of a program p is called strict if it can be formally proven that no concrete execution of p will leave the given corridor at run-time. It is called loose, if one or multiple consecutive violations of o are tolerable.

Independent from the above definition, an RRE can be classified as a *centralized* or a *distributed* enforcement technique:

Fig. 9.4 Centralized RRE

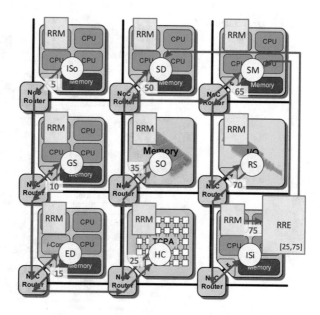

Definition 9.4 (Centralized/Distributed RRE) A Run-Time Requirement Enforcer (RRE) of a requirement $r_o(p) = [LB_o, UB_o]$ of a program p is called centralized
if a single enforcer instance is used to enforce the requirement. It is called distributed
in case multiple enforcers jointly enforce the requirement.

Figure 9.4 illustrates an example of a centralized RRE of a latency requirement
for an object detection streaming application from the area of robot vision illustrated
in Fig. 9.6. Here, the execution time of the 9 tasks (actors) of the application is
monitored by a local so-called *Run-Time Requirement Monitor (RRM)* instantiated
on each of the invaded tiles. A centralized RRE instance is also instantiated which,
in the example, receives the monitored timing information of the last actor in the
chain, i.e., Image Sink (ISi), from the RRM on the respective tile based on which
it conducts enforcement decisions. During the execution of the application, each
RRM derives the time elapsed for the execution of its local actor(s) for the current
image frame and creates a time stamp that is sent together with the processed
frame to the subsequent actor. Thus, each actor in the chain is provided with the
information about the time already elapsed for the processing of the frame by the
previous actors based on which it determines the slack available for the remainder
of the processing. Once the last actor in the chain has completed its processing, the
local RRM computes and sends a completion time stamp to the centralized RRE. In
case of a soft latency requirement, a loose RRE would react to any latency corridor
violation by adjusting the voltage/frequency of the tiles that host the time-critical
actors, i.e., SIFT Description (SD) and SIFT Matching (SM), with the goal to steer
the latency of the chain back into the corridor for subsequent image frames.

Fig. 9.5 Distributed RRE

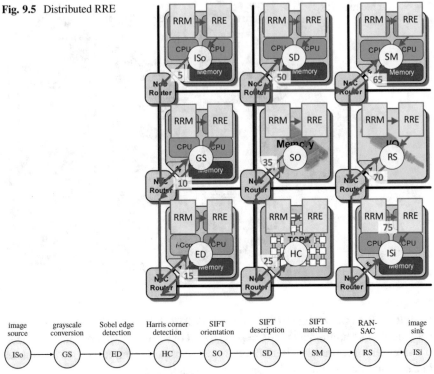

Fig. 9.6 Object detection streaming application

Figure 9.5 illustrates an example of a distributed RRE for the object detection application. Here, the overall time requirement per image frame could be partitioned into sub-corridors (or interval budgets) which are assigned to the invaded tiles. Also, in addition to the RRMs, a local RRE is instantiated per tile to enforce its assigned sub-corridor locally. Evidently, distributed RRE benefits from a simpler realization and scalability in comparison to centralized RRE. Nonetheless, centralized enforcement could better use global information to optimize secondary goals such as energy consumption, as we will show in Sect. 9.5.

9.4.1 Enforcement Automata (EA)

Although arbitrary algorithmic behavior can be envisioned for enforcement, in the following we focus on automata-based enforcement techniques, as they are simpler to generate and ideal for application of formal verification techniques for proof of correctness due to their strong formal semantics. Formal proofs are necessary

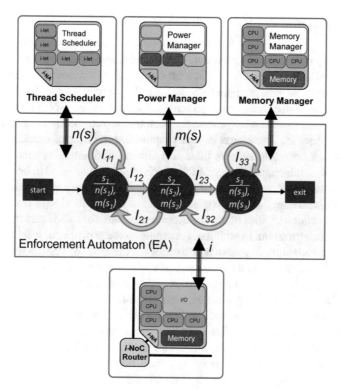

Fig. 9.7 Example of an Enforcement Automaton (EA). Depending on the input i of a program p and a current state s, the automaton takes a state transition to enforce a requirement. In the example, in state $s \in S$, the EA outputs how many cores $n(s)$ shall be powered on and in which power mode $m(s)$ (voltage, frequency) p shall be executed

particularly for enforcement of hard requirements. Figure 9.7 illustrates an example of an enforcement automaton (EA) of type Moore in which the input is a measure of the current workload i of a periodically executed program (segment or task) p, e.g., an image processing actor or kernel. In each state $s \in S$, the EA produces a vector of two outputs: the number $n(s)$ of cores to be powered on for executing the current job and the power mode $m(s)$ to be applied to the active core(s). As illustrated, the RRE acts as an interface between the application and the system software of the tile. Although in the examples provided in this paper, only the power management facility (voltage/frequency settings) and the degree of parallelism are controlled by RREs, they could in general control or restrict other system software components as well, e.g., the thread scheduler or the memory managing unit, for the enforcement of the given requirements.

9.4.2 *i-lets and e-lets*

Whereas for centralized enforcement, we assume that only one enforcer is instantiated per application program p, each task/actor of a distributedly mapped application program will be assigned its own local enforcer. In our implementation, an enforcer is implemented as a preferential thread called *e-let* in the following whereas application threads spawned for each task execution are called *i-lets*. Note that even if both are considered logically equivalent in terms of executable threads at the level of operating system, there is a notable difference between both: Even if i-lets present the application code for which requirements need to be guaranteed, they are usually not preferred by the operating system over other threads of their kind of the same application program. Whereas e-lets are always considered the preferred execution entities of the application program, they dominate the i-lets also included in this program. In addition, according to the principle of least privilege, e-lets have the capability to overrule or restrict the behavior of system-level software components including schedulers as well as cache, memory, and power managers in order to be able to enforce the properties required of their assigned i-lets. Another major difference between application i-lets and e-lets is the way they are executed. Whereas i-lets are created and executed upon each activation, e-lets are created only once at the time where an application program invades a tile of cores. They remain active not only for one iteration, but until the whole application retreats from all occupied resources. e-lets, in particular, state transitions in case the behavior is described by an EA, are triggered by incoming events, very similar to data-driven execution. In case of the following robot vision application, e.g., a state transition is triggered each time a new frame is arriving from a neighbor tile. In normal execution, the i-lets of an activated application task start after the EA has transited from the actual to the next state and have run-to-completion semantics. Whereas, e-lets may alternatively be triggered by asynchronous events, e.g., an exception from a temperature monitor.

Since an e-let is to be provided with special system privileges, including in particular the capability for immediate and low-latency response in bounded time to system events and operational state changes, it is implemented as a *kernel-level thread*. As a first-class object of the operating-system kernel, such a thread makes it possible to establish and maintain a semantic relationship between system-level and user-level code. The same applies to an i-let, but without granting the associated kernel-level thread any special privileges.

An ensemble of kernel-level threads with and without special capabilities for the control of system behavior depending on the particular requirements of an application program is managed by the kernel in the shape of a *squad*. A squad is a special unit within a *team* (a non-empty set of processes sharing a common address space and common computing resources [4]) of related kernel-level threads. This unit consists of two types of threads: on the one hand, those that make up the actual *lead* of the application program and on the other hand at least one *aide* who assists the lead threads as system mediator. From the kernel's point of view, the aide has all the capabilities to assure the lead the required system behavior in a controlled manner. In addition to be able to override or modify certain operating-system decisions, the aide is able to instantly respond to system events. In such a setting, an e-let is mapped to an aide, while the i-lets for which certain properties are enforced appear as lead.

9.5   Case Study

In this section, we present examples of enforcement techniques for strict vs. loose as well as distributed vs. centralized enforcement of timing requirements for the case study of the previously introduced object detection application depicted in Fig. 9.6. The application consists of a chain of 9 tasks (actors) processing each input image in succession: an image source (ISo) actor to read in input images periodically at a constant rate, a gray-scale (GS) conversion actor, a Sobel edge detection (ED) actor and a Harris corner (HC) detection actor to determine, respectively, edges and corners in an image, a SIFT orientation (SO) actor to achieve invariance to image rotation, a SIFT description (SD) actor to extract the *features* in an image, a SIFT matching (SM) actor to detect objects in the image based on a previously trained set of object features, and a RANSAC (RS) actor to insert the detected objects into the image which is finally sent out by an image sink (ISi) actor. As platform, we consider a NoC-based 3×3 many-core architecture as depicted in Fig. 9.1 and map the application's actors on the architecture as illustrated in Figs. 9.4 and 9.5. All evaluations presented in this section are carried out using InvadeSIM [16, 18], a high-level functional simulation framework for multi-/many-core architectures and supporting resource-aware programming.

9.5.1   Enforcement Problem Description

In the following, we assume that each image frame of the given time-critical application must be processed within a latency upper bound $UB_L = 115$ ms. Table 9.1 provides the average, standard deviation, and overall contribution of each actor's latency when processing a sequence of 9 149 images stemming from different sources of video streams when each actor is processed in isolation on a single core and running constantly at maximum frequency. As can be seen in Table 9.1, the SD and SM actors exhibit the highest degree of input-dependent variation in execution time and also the highest contribution to the overall latency. The remaining actors, on the other hand, do not exhibit a comparable execution time jitter and/or a comparable contribution to the overall application latency across the input space.

Table 9.1 Average, standard deviation, and overall contribution to the overall latency of each actor of the object detection application in Fig. 9.6 when processing a test sequence of 9 149 images and executed each in isolation on a single core running constantly at maximum frequency according to Table 9.2

Latency index	Actor						
	GS	ED	HC	SO	SD	SM	RS
Average [ms]	0,21	0,18	1,50	1,79	146,86	21,02	0,01
Std. deviation [ms]	0,09	0,08	0,64	0,80	106,15	15,04	0,03
Overall contribution	0,1 %	0,1 %	0,9 %	1,0 %	85,6 %	12,3 %	0,0 %

In the following, we present examples of RRE techniques using Dynamic Voltage and Frequency Scaling (DVFS) [3, 10, 21, 26] to enforce the global latency upper bound $UB_L = 115$ ms for the given application. Due to the small variation and overall latency contribution of all except the actors SD and SM according to Table 9.1, we dedicate a time budget of 20 ms to the other actors altogether, assuming that their cumulative latency per input image does not exceed this budget. This translates into a latency upper bound of $UB_L = 95$ ms for the SD and SM actors. For the demonstration of distributed RRE techniques, we further decide to split this bound into two individual latency upper bounds, namely $UB_L = 80$ ms for SD and $UB_L = 15$ ms for SM. Next, we present examples of loose vs. strict as well as distributed vs. centralized enforcement. As a merit of profit, we also investigate the potential energy savings of each RRE strategy in addition to evaluating its capability in enforcing the latency requirement(s).

According to Fig. 9.7, the following RRE techniques implemented as enforcement automata are privileged to adjust two control knobs prior to processing an image frame: (a) the degree of execution parallelism per actor that is adjusted by setting the number n of active cores that process the workload of each actor and (b) the power (voltage/frequency) mode m of the core(s) allocated for each actor adjusted through DVFS (for active cores) and power gating (for inactive cores). To this end, each RRE decides on a per input image basis how to distribute the workload of each actor being enforced between one and four cores available per tile according to the mappings shown in Figs. 9.4 and 9.5. At the same time, it sets the power mode of the cores of each tile to either a power-gated mode (with $f = 0$ and $V_{DD} = 0$) or 20 possible DVFS configurations (with a frequency step size of 0.2 GHz and a maximum frequency of 4 GHz) summarized in Table 9.2. For both actors under enforcement, SD and SM, we analyzed the major source of latency variation according to Table 9.1 (single core, constant maximal frequency) as stemming from the variability in the number i of *features* in each image to be processed. Therefore, this number is used as a direct indicator of the input workload to the following RRE strategies.

Table 9.2 Voltage/frequency (DVFS) modes of each core

mode m	$f(m)$ [GHz]	$V_{DD}(m)$ [V]	mode m	$f(m)$ [GHz]	$V_{DD}(m)$ [V]	mode m	$f(m)$ [GHz]	$V_{DD}(m)$ [V]	mode m	$f(m)$ [GHz]	$V_{DD}(m)$ [V]
1	0.2	0.5	6	1.2	0.91	11	2.2	1.26	16	3.2	1.58
2	0.4	0.6	7	1.4	0.98	12	2.4	1.32	17	3.4	1.65
3	0.6	0.69	8	1.6	1.05	13	2.6	1.39	18	3.6	1.71
4	0.8	0.77	9	1.8	1.12	14	2.8	1.45	19	3.8	1.78
5	1.0	0.84	10	2.0	1.19	15	3.0	1.52	20	4.0	1.84

9.5.2   Power, Latency, and Energy Model

Our investigation of enforcement strategies involves the evaluation of power consumption, execution latency, and energy demand per actor under enforcement. To evaluate the power consumption $P(m)$ of a core in power mode m, we use Eq. (9.2) in which the first summand represents the dynamic power contribution calculated based on the effective switching capacitance C_{eff} and the supply voltage $V_{DD}(m)$ and operating frequency $f(m)$ of the core in power mode m. The second summand describes the static power consumption calculated as the product of leakage current I_{leak} and supply voltage $V_{DD}(m)$.

$$P(m) = C_{eff} \cdot V_{DD}(m)^2 \cdot f(m) + I_{leak} \cdot V_{DD}(m) \qquad (9.2)$$

For the construction of proper enforcement automata, we need to know the relation between the number i of input features and the execution latency L of each actor to be enforced in dependence of the number n of cores and power mode m. Let $L(1, 1, m_{max})$ denote the latency for processing one feature on one core in power mode m_{max} (highest voltage and frequency). In the following, $L(1, 1, m_{max})$ is determined by simulatively determining the execution latency of each actor per image for a representative set of 9 149 test images that fully covers the considered input space. Subsequently, the latency per feature of an actor is determined for each image by dividing its latency by the number of features i in that image. Figure 9.8 illustrates the distribution (left) and the cumulative distribution (right) of the per-feature latency for the SD actor. Based on the obtained distribution, we then determine $L(1, 1, m_{max})$ according to the *strictness* of the latency requirement which specifies the minimum rate $s \in [0, 1]$ of requirement satisfaction that must be achieved, specified by the user. In case of (a) strict enforcement, a strictness of $s = 1$ is considered, and hence, the maximum observed per-feature latency among all images is used as $L(1, 1, m_{max})$. For (b) loose enforcement, i.e., when $s < 1$, $L(1, 1, m_{max})$ is set to the lowest per-feature latency among all images such that for $s \cdot 100\%$ of images, the latency per feature is lower than or equal to the selected $L(1, 1, m_{max})$. In Fig. 9.8 (right), this calculation corresponds to finding the lowest x-coordinate with a cumulative density of s. Having $L(1, 1, m_{max})$ determined, the following Eq. (9.3) is then used to determine the actor latency $L(i, n, m)$ based on

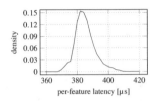 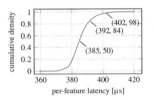

Fig. 9.8 Distribution (left) and cumulative distribution (right) of observed per-feature latency of the SD actor for a test sequence of 9 149 input images with number i of features to be processed varying between 0 and 5 513. To the right, the value of $L(1, 1, m_{max})$ is marked for requirement strictness values of $s = 0.5$, 0.84, and 0.98

the number of features i to be processed within an image, the number of cores n employed, and the power mode m selected by an RRE scheme. In Eq. (9.3), $e(n)$ denotes the parallel efficiency in dependence of the number of cores n employed for the computation with $e(n) = 1$ in the best case. In our experiments, we consider $e(n) = 1$.

$$L(i, n, m) = L(1, 1, m_{max}) \cdot \left\lceil \frac{i}{n \cdot e(n)} \right\rceil \cdot \frac{f(m_{max})}{f(m)} \qquad (9.3)$$

Note that Eq. (9.3) is a latency model specific to the SD and SM actors of our running application where $L(1, 1, m_{max})$ must be determined individually for each actor to be enforced. Moreover, Eq. (9.3) could be alternatively replaced with an elaborate many-core timing analysis, e.g., those from [2, 5–7, 15, 25], to derive tight worst-case latencies that support a variety of different resource arbitration policies and resource sharing schemes. Based on the power consumption and latency models in Eqs. (9.2) and (9.3), the energy $E(i, n, m)$ required by the actor for processing an image with i features using n cores running in power mode m is derived using Eq. (9.4).

$$E(i, n, m) = L(i, n, m) \cdot P(m) \cdot n \qquad (9.4)$$

Finally, the maximum number of features that can be processed within a given latency bound UB_L using n active cores running in power mode m can be determined using Eq. (9.5) which is derived from Eq. (9.3), considering $L(i, n, m) \leq UB_L$.

$$i_{max}(UB_L, n, m) = \left\lfloor n \cdot e(n) \cdot \left\lfloor \frac{UB_L}{L(1, 1, m_{max})} \cdot \frac{f(m)}{f(m_{max})} \right\rfloor \right\rfloor \qquad (9.5)$$

For example, with Eq. (9.5), we may compute $i_{max}(80, 4, 20)$ for the SD actor which is the highest number i of features of an input image for which a latency upper bound of $UB_L = 80\,\mathrm{ms}$ can be enforced with a strictness of $s \in [0, 1]$. For instance, for loose enforcement with $s = 0.5$, we obtain $i_{max}(80, 4, 20) = 828$, and for strict enforcement where $s = 1$, the maximum enforceable workload decreases to $i_{max}(80, 4, 20) = 760$ features.

9.5.3 Energy-Minimized Timing Enforcement

According to Fig. 9.7, RRE may involve to set, modify, or impose restrictions on typically OS-related techniques such as thread scheduling or memory management. In the following examples, we exemplify enforcement strategies for latency enforcement of individual actor executions or complete applications by varying the number $n \in [1, 4]$ of cores (parallelism) and the power mode $m \in [1, 20]$ configuration for each actor execution. As, in general, multiple ways and settings for n and m might be feasible to enforce a requirement, the question becomes which requirement-adhering constellation the enforcer selects at run-time. Often, this freedom of choice may be exploited by optimizing one or more (secondary) objectives in addition to satisfying the given requirement. In the following, we consider energy demand as an objective to be minimized.[2] Given a latency requirement UB_L and the RRE decision space of $n \in [1, 4]$ and $m \in [1, 20]$, design space exploration can be conducted per actor (or a set of actors) to derive, e.g., in our running example for the SD actor, the maximum number i_{max} of features that can be processed under each choice of (n, m) while respecting the latency requirement. Taking the SD actor with a latency upper bound $UB_L = 80$ ms as an example, Fig. 9.9 illustrates the maximum workload i_{max} and the respective energy demand for each of the 80 possible (n, m) configurations

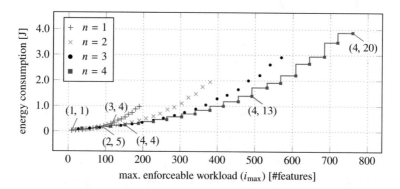

Fig. 9.9 Maximum enforceable workload i_{max} and energy demand of the SD actor under the variation of the number $n \in [1, 4]$ of active cores and their power mode $m \in [1, 20]$ for a hard $(s = 1)$ latency bound of $UB_L = 80$ ms. Pareto-optimal (n, m) configurations are connected by a red line. For an exemplary subset of them, the Pareto-optimal configuration (n, m) is also annotated

---

[2]Other objectives for choice of settings could be to activate the least number n of cores for increasing aspects of long-term reliability.

derived using Eqs. (9.5) and (9.4), respectively, in case of strict enforcement ($s = 1$). The red line designates Pareto-optimal (n, m) configurations.

Based on such a design space exploration and the Pareto front of (n, m) configurations derived thereby, an energy-minimizing enforcement automaton may be systematically constructed in which prior to each execution of the SD actor, the enforcement automaton selects a state (Pareto-optimal (n, m) configuration) that is energy-minimal while satisfying the latency requirement in case input i is enforceable, thus if $i \leq i_{\max}(UB_L, n, m)$. For the example in Fig. 9.9, the enforcement automaton has 31 states, each corresponding to one of the 31 Pareto-optimal (n, m) configurations and the maximum enforceable workload $i_{\max}$ associated with that configuration. Here, the state selection is steered solely by the number i of features in the image to be processed by the SD actor.[3] For instance, for images with $i \leq 9$ features, $n = 1$ and $m = 1$ minimizes the energy demand of the SD actor without violating the given latency requirement $UB_L = 80$ ms. For input images with $142 \leq i \leq 152$ features, an energy-minimal and requirement-adhering execution can be realized only if $n = 4$ cores are used for SD in parallel and power mode $m = 4$. Finally, a strict enforcement becomes impossible if $i > 760$, even using the configuration with the highest compute power, i.e., $n = 4$ and $m = 20$. For non-enforceable inputs, the enforcer needs to either throw an exception, stop processing (drop) the image, or process only as much as the latency bound allows to be processed. In Sect. 9.5.6, we propose a number of exception handling techniques under the topic of range extension. Before that, we first present techniques for distributed enforcement where each actor is individually enforced. Subsequently, we present also an example of centralized enforcement in which a more global view of the system state can be obtained by a centralized RRE instance that can take decisions affecting multiple actors and resources.

9.5.4 Distributed Enforcement

Figure 9.10 shows the resulting automatically generated energy-minimizing enforcement automata for a distributed enforcement strategy of the two individual actors SD and SM with latency upper bounds 80 ms and 15 ms, respectively. The EAs for selecting the energy-minimizing (n, m) configurations obtained through the previously presented design space exploration are implemented as lightweight lookup tables for each actor. At run-time, once an image is ready to be processed, the number i of features in it becomes known. Prior to processing an image, the RRE (e-let) retrieves the energy-minimizing (n, m) configuration corresponding to

---

[3]Note that in this example, the RRE could also be represented by a function table rather than an FSM, as the selection of state is only dependent on the input. More general cases such as restricting the allowed settings in each state to allow only step-wise increase or decrease of DVFS modes can be constructed.

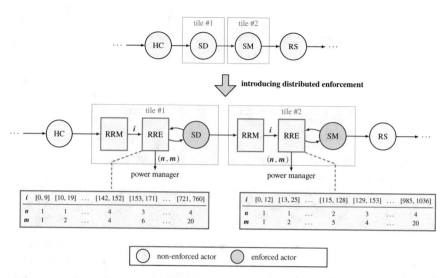

Fig. 9.10 Implementation of distributed RRE using pre-explored energy-optimal parallelism degree and DVFS settings (n, m) for SD and SM actors with hard $(s = 1)$ latency upper bounds of $UB_L = 80$ ms and $UB_L = 15$ ms, respectively

i features from the table and instructs the power manager to use these settings. As shown in Fig. 9.10, the integration of enforcers may be achieved at the level of actor graphs as a model transformation by inserting the enforcer as an actor in front of each actor to be enforced, such that for each image to be processed, the energy-minimizing (n, m) configuration is set prior to execution of the image, and the configuration stays constant over the duration of processing this image. Employing the above enforcement strategy, the run-time manager is not compelled to run the enforced actors constantly with the maximum number $n = 4$ of cores and in the highest power mode $m = 20$ to guarantee the satisfaction of latency constraints in the presence of input variations, unless $i \geq 721$ for SD or $i \geq 985$ for SM. Also note that the given latency bounds cannot be strictly enforced for a feature count $i > 760$ for SD and $i > 1\,036$ for SM. Thus, the maximum workload that can be strictly enforced by both actors is limited to $i = 760$ features.

The histograms of observable latencies of the SD and SM actors (a) without enforcement ($n = 4$ and $m = 20$) and (b) with enforcement considering hard ($s = 1$) latency upper bounds of $UB_L = 80$ ms and 15 ms for SD and SM, respectively, are illustrated in Fig. 9.11. As shown in the plots, the RREs choose a power mode that maximizes energy savings while satisfying the given latency upper bound of each actor under enforcement. For a variety of requirement strictness levels, Table 9.3 finally presents the average dynamic energy consumption and the achieved dynamic energy savings of the SD and SM actors compared to the non-enforced scenario with $n = 4$ and $m = 20$. As can be seen, in case of *loose enforcement*, i.e., a strictness $s < 1$, the RRE achieves between 38.3 % and 41.2 % dynamic energy savings per enforced actor (respectively, between 39.3 % and 40.8 % collectively

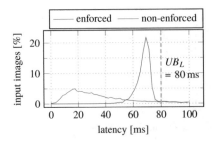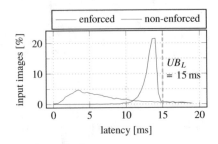

Fig. 9.11 Latency distribution for the SD (left) and SM (right) actors. The enforced case corresponds to the energy-minimized enforcement for hard ($s = 1$) latency bounds of $UB_L = 80$ ms for SD and $UB_L = 15$ ms for SM. The non-enforced case corresponds to a fixed setting of $n = 4$ and $m = 20$ per actor

for the two actors) while satisfying latency upper bounds of 80 ms and 15 ms for the SD and SM actors, respectively. In case of *strict enforcement* which corresponds to a requirement satisfaction rate of $s = 1$, the RRE still is able to achieve dynamic energy savings of 37.6% for SD and 37.2% for SM (respectively, 37.6% collectively for the two actors) while guaranteeing that the given latency upper bound for each actor will never be violated. Evidently, this guarantee holds only for enforceable input images, i.e., those with $i \leq 760$ features for the SD actor and $i \leq 1\,036$ for SM (see the RRE tables in Fig. 9.10). In Sect. 9.5.6, we discuss approaches that can be employed to enable the enforcement of latency requirements for inputs which are not enforceable merely using the given RRE control knobs. Finally, when analyzing the overall energy consumption of all actors per input frame, we obtain an overall dynamic energy reduction of 33.8% in case of strict enforcement ($s = 1$) and between 35.4% and 36.8% in case of loose enforcement ($s < 1$) for the whole application, even though only two out of 9 actors are enforced. Noteworthy, the additional execution time and energy consumption of the RREs themselves can be neglected as these are implemented by simple table lookups.

9.5.5 Centralized Enforcement

In this section, we consider the combined enforcement of the SD and SM actors using centralized enforcement. As depicted in Fig. 9.12, a single instance of an RRE is now enforcing the overall hard ($s = 1$) latency upper bound of $UB_L = 80 + 15 = 95$ ms for both SD and SM actors collectively. Similar to the distributed case, the energy-minimizing (n, m) configurations which are required for the construction of the RRE are obtained through a previously presented design space exploration, but now considering a unified latency upper bound $UB_L = 95$ ms for the execution of both SD and SM actors. Note that considering a compound

Table 9.3 Average dynamic energy consumption and savings per image for the SD and SM actors through distributed enforcement in dependence of requirement strictness defined as the minimum acceptable requirement satisfaction rate

Requirement strictness		SD actor ($UB_L = 80$ ms)			SM actor ($UB_L = 15$ ms)			Energy savings	
norm. dist. index	min. sat. rate (s)	$L(1, 1, m_{max})$ [μs]	Avg. energy [mJ]	Energy savings	$L(1, 1, m_{max})$ [μs]	Avg. energy [mJ]	Energy savings	SD+SM	Overall
Median	50 %	385,4	131,1	41,2 %	55,4	28,9	38,9 %	40,8 %	36,8 %
avg+1·σ	84,1 %	392,0	132,9	40,4 %	55,6	29,0	38,7 %	40,1 %	36,1 %
avg+2·σ	97,7 %	402,3	135,0	39,5 %	55,9	29,1	38,3 %	39,3 %	35,4 %
Maximum	100 %	420,7	139,2	37,6 %	57,8	29,6	37,2 %	37,6 %	33,8 %
Without enforcement	–	–	223,2	(ref)	–	47,2	(ref)	(ref)	(ref)

The energy consumption without enforcement ($n = 4$, $m = 20$) serves as a baseline

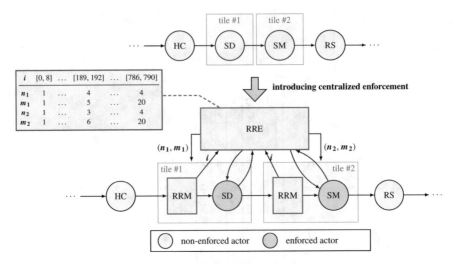

Fig. 9.12 Implementation of a centralized RRE using pre-explored energy-optimal parallelism degree and DVFS settings (n, m) for SD and SM actors with a hard $(s = 1)$ latency upper bound of $UB_L = 95$ ms for both actors collectively

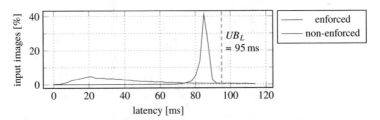

Fig. 9.13 Latency distribution of the SD and SM actors when collectively enforced for a hard $(s = 1)$ compound latency bound of $UB_L = 95$ ms. The enforced scenario is realized using energy-minimized enforcement, and the non-enforced scenario corresponds to a fixed configuration of $n = 4$ and $m = 20$ per actor

latency bound for both actors enables enforcing this bound for images with up to $i = 790$ features.

The histogram of observable collective latency of the SD and SM actors (a) without enforcement ($n = 4$ and $m = 20$ for both actors) and (b) with enforcement considering a hard latency upper bound, i.e., for a requirement strictness of $s = 1$, is illustrated in Fig. 9.13. As shown in the plots, the RRE assigns the number n of active cores and their power mode m for each actor under enforcement to maximize energy savings while satisfying the given latency upper bound of $UB_L = 95$ ms collective for both actors. For a variety of requirement strictness levels, Table 9.4 finally presents the average dynamic energy consumption and the achieved dynamic energy savings of the two enforced actors using centralized enforcement compared to the non-enforced scenario with $n = 4$ and $m = 20$. As can be seen, the RRE achieves in case of *loose enforcement*, i.e., strictness $s < 1$, between 39.7 %

Table 9.4 Average dynamic energy consumption and savings per image for the SD and SM actors through centralized enforcement for a compound latency bound of $UB_L = 95$ ms in dependence of requirement strictness defined as the minimum acceptable requirement satisfaction rate

Requirement strictness		SD actor			SM actor			Energy savings	
norm. dist. index	min. sat. rate (s)	$L(1, 1, m_{max})$ [μs]	Avg. energy [mJ]	Energy savings	$L(1, 1, m_{max})$ [μs]	Avg. energy [mJ]	Energy savings	SD+SM	Overall
median	50 %	385,4	129,9	41,8 %	55,4	26,7	43,6 %	42,1 %	37,9 %
avg+1·σ	84,1 %	392,0	131,3	41,2 %	55,6	27,0	42,9 %	41,5 %	37,3 %
avg+2·σ	97,7 %	402,3	135,3	39,4 %	55,9	27,9	40,9 %	39,7 %	35,7 %
maximum	100 %	420,7	137,2	38,5 %	57,8	28,2	40,2 %	38,8 %	35,0 %
Without enforcement	–	–	223,2	(ref)	–	47,2	(ref)	(ref)	(ref)

The energy consumption without enforcement ($n = 4$, $m = 20$) serves as a baseline

and 42.1 % dynamic energy savings collectively for the two enforced actors while satisfying a compound latency upper bound of 95 ms. In case of *strict enforcement* ($s = 1$), the RRE still is able to achieve dynamic energy savings of 38.8% collectively for the two actors while guaranteeing that the given latency upper bound $UB_L = 95$ ms will never be violated. Finally, when analyzing the overall energy consumption of all actors per input frame, we obtain a dynamic energy reduction of 35 % in case of strict enforcement ($s = 1$) and between 35.7 % and 37.9 % in case of loose enforcement ($s < 1$), even though only two out of 9 actors are enforced. In summary, compared to distributed enforcement, the centralized scheme is able to even save slightly more dynamic energy while enforcing a higher workload.

9.5.6 Lower Latency Bound Enforcement and Range Extenders

In certain cases, a latency requirement may introduce—in addition to an upper bound UB_L—also a lower bound, LB_L, thus, demanding the enforcement of a latency corridor. Such a lower latency bound could be enforced by means of, e.g., a simple timer (counter) that measures the time elapsed from the beginning of the current execution of the actor(s) under enforcement. The transmission of the produced result(s) to the next actor(s) could then be simply delayed to the time the timer indicates that the time interval of LB_L has passed, see Fig. 9.14. More

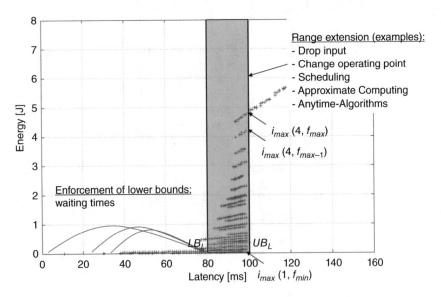

Fig. 9.14 Examples of range extenders and enforcement of lower latency bounds LB_L and thus latency corridors

difficult and also diverse in the space of possible solutions, however, is the question of how to deal with non-enforceable inputs. In case of our running distributed object detection application, our test image sequences on purpose contained images with more features i than for which the given latency upper bound can be enforced with only $n = 4$ cores in highest power mode $m = 20$. In case of strict enforcement ($s = 1$) corresponding to hard real-time requirements, not even a single violation of a latency upper bound is tolerable. Hence, there must be techniques to avoid such violations per construction, if a non-enforceable input is observed. This is a matter of current research. We therefore briefly outline a few techniques how to deal with these cases: input omission (dropping), approximate computing to trade off processing speed with result accuracy (if applicable), revision of scheduling decisions, over-allocation of resources, or a dynamic reconfiguration between different mappings at run-time (change of operating point [14]), see also Fig. 9.14.

9.6 Conclusions

In this paper, we presented a formalization, classification, and the practice of a class of run-time techniques subsumed under the term of Run-Time Requirement Enforcement (RRE) that make the system management software of an MPSoC platform become the advocate of a parallel application program instead of both acting independently with the goal to provide means for the satisfaction of given non-functional requirements of parallel program execution such as performance (latency, throughput), power or energy consumption, or reliability. The non-functional requirements can thereby be expressed by interval ranges and specified over the application program as a whole, e.g., when specified by an actor graph. Alternatively, requirements can be specified for individual actors/tasks or threads, or even segments thereof. The goal of RRE is to enforce the satisfaction of these requirements at run-time. It has been shown by introductory examples on latency enforcement of a distributed object detection application that enforcers may be generated through profiling and the creation of high priority system-level threads called *e-lets* that are formally described in behavior by an enforcement automaton each. These e-lets proactively control the system resources claimed by an application program in view of observed workload variation. First, based on the assignment of exclusive resources to periodic workload such as streaming applications, composability is created that is necessary to allow for a static and independent analysis of each application running on a given MPSoC platform. This enables us to statically analyze non-functional properties of applications or parts thereof and define RRE techniques to control requirements dynamically. For a distributed object detection application as an example, it has been shown that the variability of non-functional execution properties can be greatly reduced in dependence of the level of strictness that shall be fulfilled for each requirement. Moreover, it has been shown that RRE techniques can be either implemented in a centralized or distributed manner. In the future, we want to look at how to decompose requirement corridors for distributed enforcement and study the control overheads of centralized enforcement. Finally, techniques for simultaneous enforcement of multiple non-functional requirements need to be investigated, as here, not only the input (workload) variation as considered in this seminal paper, but also the shared system state must be taken into account once multiple RREs are at work.

Acknowledgments This paper is dedicated to Peter Marwedel on behalf of his 70th birthday in recognition of his lifetime achievements in the area of design automation for embedded systems. Special thanks also to Zhai Ming for several experiments conducted for this paper. Finally, we would like to acknowledge the Deutsche Forschungsgemeinschaft (DFG, German Research Foundation)—Project Number 146371743—TRR 89 Invasive Computing that funded our work.

References

1. B. Akesson, et al., Composability and predictability for independent application development, verification, and execution, in *Multiprocessor System-on-Chip* (Springer, Berlin, 2011), pp. 25–56
2. S. Altmeyer, et al., A generic and compositional framework for multicore response time analysis, in *Proceeding of RTNS* (ACM, New York, 2015), pp. 129–138
3. D. Angioletti, et al., A runtime resource management policy for OpenCL workloads on heterogeneous multicores, in *Proceeding of DATE* (IEEE/ACM, New York, 2019), pp. 1385–1390
4. D.R. Cheriton, et al., Thoth, a portable real-time operating system. Commun ACM **22**(2), 105–115 (1979)
5. R.I. Davis, et al., An extensible framework for multicore response time analysis. Real-Time Syst. **54**(3), 1–55 (2017)
6. G. Giannopoulou, et al., Timed model checking with abstractions: towards worst-case response time analysis in resource-sharing manycore systems, in *Proceedings of the International Conference Embedded Software* (ACM, New York, 2012), pp. 63–72
7. G. Giannopoulou, et al., Mixed-criticality scheduling on cluster-based manycores with shared communication and storage resources. Real-Time Syst. **52**(4), 399–449 (2016)
8. A. Hansson, et al., CoMPSoC: a template for composable and predictable multi-processor system on chips. ACM TODAES **14**(1), 2 (2009)
9. C. Imes, et al., POET: a portable approach to minimizing energy under soft real-time constraints, in *Proceeding of RTAS* (IEEE, Silver Spring, 2015), pp. 75–86
10. A. Kanduri, et al., Approximation-aware coordinated power/performance management for heterogeneous multi-cores, in *Proceeding of DAC* (IEEE/ACM, New York, 2018), pp. 1–6
11. P.N. Khanh, et al., Incorporating energy and throughput awareness in design space exploration and run-time mapping for heterogeneous MPSoCs, in *Proceeding of DSD* (IEEE, Silver Spring, 2013), pp. 513–521
12. H. Kopetz, *Real-time Systems: Design Principles for Distributed Embedded Applications*, 2 edn. (Springer, Berlin, 2011)
13. S. Pinisetty, et al., Runtime enforcement of reactive systems using synchronous enforcers, in *Proceeding of ACM SIGSOFT International SPIN Symposium Model Checking of Software* (2017), pp. 80–89
14. B. Pourmohseni, et al., Hard real-time application mapping reconfiguration for NoC-based many-core systems. Real-Time Syst. **55**(2), 1–37 (2019)
15. B. Pourmohseni, et al., Isolation-aware timing analysis and design space exploration for predictable and composable many-core systems, in *Proceeding of ECRTS* (2019)

16. S. Roloff, et al., Execution-driven parallel simulation of PGAS applications on heterogeneous tiled architectures, in *Proceeding of DAC* (IEEE/ACM, New York, 2015), pp. 1–6
17. S. Roloff, et al., ActorX10: an actor library for X10, in *Proceeding of ACM SIGPLAN Workshop on X10* (ACM, New York, 2016), pp. 24–29
18. S. Roloff, et al., *Modeling and Simulation of Invasive Applications and Architectures* (Springer, Berlin, 2019)
19. T. Schwarzer, et al., Symmetry-eliminating design space exploration for hybrid application mapping on many-core architectures. IEEE TCAD **37**(2), 297–310 (2018)
20. A.K. Singh, et al., Accelerating throughput-aware runtime mapping for heterogeneous MPSoCs. ACM TODAES **18**(1), 9:1–9:29 (2013)
21. A.K. Singh, et al., Energy optimization by exploiting execution slacks in streaming applications on multiprocessor systems, in *Proceeding of DAC* (IEEE/ACM, New York, 2013), p. 115
22. J. Teich, et al., *Invasive Computing: An Overview* (Springer, New York, 2011)
23. J. Teich, et al., Language and compilation of parallel programs for *-predictable MPSoC execution using invasive computing, in *Proceeding of MCSOC* (IEEE, Silver Spring, 2016)
24. A. Weichslgartner, et al., DAARM: design-time application analysis and run-time mapping for predictable execution in many-core systems, in *Proceeding of CODES+ISSS* (IEEE/ACM, New York, 2014), pp. 1–10
25. A. Weichslgartner, et al., *Invasive Computing for Mapping Parallel Programs to Many-Core Architectures* (Springer, Berlin, 2018)
26. Z. Zhu, et al., Energy minimization for multi-core platforms through DVFS and VR phase scaling with comprehensive convex model, in *IEEE Transactions on Computer-Aided Design of Integrated Circuits and Systems* (2019)

Chapter 10
Compilation for Real-Time Systems a Decade After PREDATOR

Heiko Falk, Shashank Jadhav, Arno Luppold, Kateryna Muts,
Dominic Oehlert, Nina Piontek, and Mikko Roth

10.1 Introduction

PREDATOR was a collaborative research project running from February 2008 until January 2011 that was funded by the European 7th Framework Programme under the lead of Reinhard Wilhelm, Saarland University, Germany. It was concerned *"with embedded systems that are characterized by efficiency requirements on the one hand and worst-case constraints on the other. [...] Embedded systems with critical constraints need off-line guarantees for the satisfaction of these constraints. Unfortunately, it can be observed that in computer system design, the gap between average-case and worst-case behavior increases rapidly. This entails a decreasing precision of performance analysis results."* Therefore, PREDATOR proposed *"a new research and design discipline that looks at predictability and efficiency in a synergistic manner and that involves all levels of abstraction and implementation in embedded system design"* [6, 34].

These different abstraction levels were reflected by the project's scientific work packages. WP1 (led by Luca Benini, University of Bologna, Italy) dealt with predictable and efficient hardware architectures. Both functional and power models of a predictable architecture were developed and their sensitivity to architectural parameters that influence predictability and their costs were analyzed. WP2 (lead: Peter Marwedel, University of Dortmund, Germany) focused on compiler and code generation techniques for a single application task. Here, optimizations that are aware of hard real-time constraints and of Worst-Case Execution Times (WCET) were proposed; multi-objective trade-offs between real-time guarantees and energy consumption or code size were envisioned. WP3 was led by Giorgio Buttazzo

H. Falk (✉) · S. Jadhav · A. Luppold · K. Muts · D. Oehlert · N. Piontek · M. Roth
Institute of Embedded Systems, Hamburg University of Technology (TUHH), Hamburg, Germany
e-mail: Heiko.Falk@tuhh.de

© The Author(s) 2021
J.-J. Chen (ed.), *A Journey of Embedded and Cyber-Physical Systems*,
https://doi.org/10.1007/978-3-030-47487-4_10

(Scuola Superiore Sant'Anna, Pisa, Italy) and targeted the coordination of multiple application tasks. Off-line and online coordination techniques were investigated such that guarantees on tasks' response times were derived under simultaneous optimization of resource usage. In the context of WP4 (lead: Lothar Thiele, ETH Zürich, Switzerland), distributed embedded systems were considered and the modular analysis of Multi-Processor Systems on Chip (MPSoC) with respect to performance and predictability was investigated. Finally, cross-layer aspects of the design and analysis of predictable and efficiency were considered in WP5 (lead: Reinhard Wilhelm).

Overall, PREDATOR was a high-quality collaborative effort that produced many seminal results in the field of designing predictable and efficient hardware and software architectures. On the occasion of Peter Marwedel's 70th anniversary, this article surveys the results in the area of compilers for real-time systems that have been achieved under his leadership within PREDATOR. The foundational character of this project is highlighted by providing an overview over code optimizations and analyses proposed in the past decade since PREDATOR was executed. These recent works directly base on challenges identified during and on results produced by PREDATOR.

Section 10.2 puts the state-of-the-art in compilation for real-time systems by the end of PREDATOR in a nutshell. Recent developments that integrate task coordination into compiler optimizations are described in Sect. 10.3. The combination of system-level analysis and code generation techniques for parallel multi-core systems is the subject of Sect. 10.4. Section 10.5 discusses multi-objective compiler optimizations that are able to adhere to real-time constraints, and Sect. 10.6 concludes this article and provides an outlook over future work.

10.2 Challenges and State-of-the-Art in WCET-Aware Compilation During PREDATOR

A program's WCET stands for its maximal possible execution time, irrespective of possible input data and of initial states of the hardware architecture. For the design of hard real-time systems, the WCET is a critical design parameter, since it allows to reason about whether a program always meets its deadline or not. However, the exact computation of a program's WCET is infeasible in general so that conservative WCET estimates are used instead. In the domain of hard real-time systems, such WCET estimates are usually produced by static timing analysis tools, e.g., aiT [1]. During PREDATOR's single-task activities carried out within Work Package WP2, such a timing analyzer was tightly integrated into a compiler framework. This allowed the compiler to perform WCET analyses in a fully automated fashion during code generation. The WCET data gathered this way constitutes a precise worst-case timing model inside the compiler which contrasts sharply with standard compilers that focus on average-case scenarios and that do not feature any timing models at all. The resulting WCET-aware C Compiler WCC [8] finally exploits this precise timing model in dedicated WCET-aware, single-task code optimizations.

However, modern real-time systems do not consist of only a single task—they are multi-task systems instead where tasks are preempted and scheduled according to an operating system's scheduling policy. Thus, the design of a timing predictable multi-task system includes the consideration of all tasks' end-to-end latencies including blocking times due to preemptions, i.e., the tasks' Worst-Case Response Times (WCRT). Based on the tasks' WCRTs, a subsequent schedulability analysis can be used to determine whether all tasks definitely meet their respective deadlines. Since WCETs are characterized by the behavior of machine code for a given processor architecture, and since WCRTs and schedulability analyses rely on given WCET values and mostly depend on task-level scheduling properties, there is a natural link between compilers and operating systems: the former generate the machine code that the latter have to schedule. This link was already identified during PREDATOR:

Challenge #1

"The compiler [...] will apply optimizations not for each individual task in isolation, but will consider all tasks of the entire system in a holistic view. Furthermore, it is planned to take the individual scheduling policies [...] into account" [5].

Plazar et al. proposed a software-based cache partitioning for real-time multi-task systems [29]. Cache partitioning is able to make the behavior of instruction caches more predictable, since each task of a system is assigned to a unique cache partition. The tasks in such a system can only evict cache lines residing in the partition they are assigned to. As a consequence, multiple tasks do not interfere with each other any longer w.r.t. the cache during context switches. This allows to apply static WCET analysis for each individual task of the system in isolation. The overall WCET of a multi-task system using partitioned caches is then composed of the WCETs of the single tasks given a certain partition size, plus the overhead required for scheduling and context switching. Until the completion of PREDATOR, an integration of schedulability analyses and a consideration of individual scheduling policies during compilation could not be realized due to a shortage of time.

In the context of performance analysis for massively parallel multi-core architectures, PREDATOR proposed a modular approach where a WCET analysis is performed for each application per individual processor core in isolation. By exploiting how often each core accesses the shared bus that connects all cores in a given MPSoC architecture, the additional timing interference that each processor core exhibits due to temporarily blocked bus accesses is estimated. According to PREDATOR's design rules for predictable architectures [38], TDMA-arbitrated shared buses were considered during modular performance analysis. In the end, upper timing bounds of all applications running on all processor cores are derived in a modular fashion which allows to reason about schedulability for such parallel multi-core systems [30].

Various execution models for the applications running on such an MPSoC architecture were considered. In the so-called Dedicated Access Model, applications are structured into three distinct phases: acquisition, execution, and replication. Only during the first and the latter, a task is allowed to access the shared bus in order to fetch input data or to write back computed results, resp. Since the main execution phase of a task is free of shared bus accesses, it cannot suffer from delays induced by other cores which allows for a very precise timing analysis. In the General Access Model, accesses to the shared bus can happen anytime during acquisition, replication, and execution. Thus, a timing analysis becomes more pessimistic here [31].

As the precision of timing analysis for MPSoCs thus strongly depends on the execution behavior of tasks, mechanisms enforcing well-suited and predictable access patterns to shared buses would be advantageous.

Challenge #2

"One new possibility to reduce the effect of (timing) interactions [. . .] is the use of traffic shapers. It is an open problem to include these units into a system-wide performance analysis that considers computation and communication resources" [5].

However, PREDATOR did not come up with approaches addressing this challenge.

PREDATOR explicitly considered the trade-off between predictability where hard constraints on a system's resource usage must be met versus the efficiency of a system in the average case.

Challenge #3

"We will develop models capturing various optimization objectives within the compiler, e.g. code size or energy dissipation [. . .]. Novel optimization strategies are designed in order to minimize an objective other than WCET, under simultaneous adherence to real-time constraints" [5].

Since the WCC compiler featured a detailed WCET timing model right from the project start, and since modeling code size at the assembly code level is trivial from a compiler's point of view, it was obvious to consider trade-offs between these two objectives in the beginning. For this purpose, simple heuristics for the optimization Procedure Cloning were proposed where WCETs were minimized as long as the resulting code sizes did not exceed a user-provided threshold [19]. Later, WCC was coupled with an instruction set simulator allowing to perform dynamic profiling during compilation. Furthermore, data from an instruction-level energy model [32] was also integrated. This way, the compiler was able to simultaneously model WCET, code size, ACET, and energy consumption of generated machine code.

These models were used to determine Pareto-optimal sequences of compiler optimizations. It is a well-known problem that the order in which a compiler applies its optimizations can have a significant impact on the quality of the finally generated code. In the context of PREDATOR, a stochastic evolutionary multi-objective algorithm [20, 21] found optimization sequences that trade pairs of objectives, i.e., ⟨WCET, ACET⟩ and ⟨WCET, code size⟩, resp. True multi-objective code optimizations that inherently model and consider different criteria at the same time during code generation have, however, not been investigated in depth during PREDATOR.

10.3   Integration of Task Coordination into WCET-Aware Compilation

Many architectures are equipped with fully software-controllable secondary memories. These are memories that are tightly integrated with the processor to achieve the best possible performance. These Scratchpad Memories (SPMs) can be accessed directly and are therefore in general well-suited for optimizations regarding energy consumption and execution times.

SPMs turned out to be ideal for WCET-centric optimizations, since their timing is fully predictable. The WCC compiler exploits SPMs for WCET minimization by placing assorted parts of a program into a scratchpad memory. During PREDATOR, an Integer-Linear Program (ILP) originally proposed by Suhendra et al. [33] was extended towards a single-task SPM allocation where binary decision variables x_i are used per basic block b_i. b_i is moved from main memory onto the scratchpad memory if x_i equals 1. The overall goal of this ILP is to find an assignment of values to the variables x_i such that the resulting SPM allocation leads to the minimal WCET of the whole task. Constraints are added to the ILP that model the task's internal program structure. For each basic block b_i and each successor b_{succ} of b_i in the task's Control Flow Graph (CFG), a constraint is set up bounding the WCET c_i of b_i:

$$c_i \geq c_{succ} + cost_{i,main\_mem} - gain_i * x_i \tag{10.1}$$

This constraint states that the WCET c_i of a path starting in b_i must be larger than the WCET c_{succ} of any of the successors of b_i, plus the contribution of b_i to the WCET itself with b_i located in main memory ($cost_{i,main\_mem}$), minus the potential gain when moving b_i from main memory onto the scratchpad memory. Additional constraints in the ILP model loops and function calls. The limited available capacity of the SPM is considered as well as the additional overhead due to long-distance jumps from the main memory to the SPM or back. In the end, the WCET of an entire task is represented in the ILP model by a variable $c_{\texttt{main}}^{entry}$ which models the WCET of the path starting at the task's entry point, i.e., at its `main` function [7].

This basic ILP model turned out to be very powerful and flexible so that it served as the basis for the optimization of multi-task systems. For this purpose, all tasks of a multi-task application were modeled in the ILP as described above. As a consequence, the ILP variables c_j associated with the entry points of the tasks τ_j describe a safe upper bound of the tasks' WCETs. An early work [22] towards PREDATOR's Challenge #1 on optimization of multi-task systems under consideration of scheduling policies integrated Joseph's schedulability analysis [15] into this multi-task ILP.

For priority-based scheduling, a task τ_j's WCRT r_j is the maximum possible time interval between the activation of a task and its end, including penalties due to preemptions by higher-priority tasks. The tasks' WCRTs are computed as follows:

$$r_j = c_j + \sum_{h=0}^{j-1} \left\lceil \frac{r_j}{T_h} \right\rceil * c_h \qquad (10.2)$$

Eq. (10.2) accumulates the net WCET c_j of task τ_j and the penalties due to tasks $\tau_0, \ldots, \tau_{j-1}$ having higher priority than τ_j. Each such high-priority task τ_h preempts τ_j a total of $\left\lceil \frac{r_j}{T_h} \right\rceil$ times where T_h denotes a task's period. For each preemption of τ_j by τ_h, the higher-priority task's WCET c_h is considered.

However, it is not straightforward to integrate Eq. (10.2) into an optimization's ILP, since both the WCETs c_h and the WCRTs r_j are ILP variables so that the multiplication of $\left\lceil \frac{r_j}{T_h} \right\rceil$ by c_h is infeasible. In order to solve this problem, an integer variable $p_{j,h}$ is added to the ILP for every combination of low- and high-priority tasks τ_j and τ_h, resp. $p_{j,h}$ denotes the timing penalty that is added to τ_j's WCRT due to preemptions by τ_h. Using these variables, Eq. (10.2) can be rewritten to:

$$r_j = c_j + \sum_{h=0}^{j-1} p_{j,h} \qquad (10.3)$$

In order to model $p_{j,h}$, the following linearization scheme is applied: If r_j is lower than or equal to τ_h's period T_h, τ_j can be preempted at most once by τ_h, thus leading to $p_{j,h} = 1 * c_h$. If r_j is greater than T_h but lower than or equal to $2 * T_h$, $p_{j,h} = 2 * c_h$ results, etc. In general, it has been proven that

Theorem 10.1 *If τ_j is preempted at least N times by τ_h, then $p_{j,h} \geq (N+1) * c_h$ must hold.*

Such so-called conditional constraints can efficiently be translated into ILP in Eq. [25]. A natural upper bound for the number N of preemptions of τ_j by τ_h is $\left\lceil \frac{D_j}{T_h} \right\rceil$ where D_j denotes task τ_j's deadline. Thus, the conditional constraints from Theorem 10.1 are added to the ILP for all values of N with $0 \leq N \leq \left\lceil \frac{D_j}{T_h} \right\rceil - 1$ and for all pairs of low- and high-priority tasks τ_j and τ_h, resp. Finally, the schedulability of the entire multi-task set is ensured during this ILP-based optimization by adding constraints

$$r_j \leq D_j \tag{10.4}$$

such that the WCRT of each task τ_j must be within its respective deadline.

While this work is a first step towards schedulability-aware compiler optimization, it suffers from a couple of limitations: First, the task model only supports fixed-priority scheduling and periodic tasks. Second, preemption costs due to the execution of an actual scheduler and context switching overheads are not considered. Finally, the number of constraints of the ILP proposed in [22] grows quadratically with the size of the task set, and it depends on the actual values for tasks' deadlines and periods.

The consideration of Liu and Layland's schedulability test [18] helped to overcome the limitation to fixed priorities:

$$u = \sum_j \frac{c_j}{T_j} \leq 1 \tag{10.5}$$

Eq. (10.5) states that a system that is scheduled with dynamic priorities using Earliest Deadline First (EDF) is schedulable if and only if the system load u is less than or equal to 1. Due to the already linear nature of Eq. (10.5), it is easy to integrate this schedulability test into an ILP [22].

The relaxation of strictly periodic task sets required to use an event-based task model supporting arbitrary task activation patterns and deadlines [24]. For this purpose, the ILP described above has been extended by support for density and interval functions η and ϵ, resp., as originally proposed by Gresser [10] and later taken up by Albers et al. [2]. In this approach, an arbitrary kind of task activation pattern can be characterized by the density function η that denotes the maximum number of events (i.e., task activations) in some time interval Δt. The interval function ϵ models the inverse behavior and returns the minimal time interval Δt in which n tasks are activated. This task model provides a high flexibility so that periodical multi-task systems, periodical systems with jitter or bursts, or systems with fully arbitrary task activations can be modeled in the optimization's ILP.

The consideration of an actual scheduler's overhead for context switching can be added to the ILP-based framework described above by introducing an implicit task τ_0 with the highest priority into the multi-task system. τ_0 represents the periodically executed scheduler, and by considering an actual scheduler's WCET c_0 and its period T_0, it can smoothly be integrated into the optimization framework [23].

As an alternative to Joseph's schedulability test (cf. Eq. (10.2)), Baruah proposed the so-called processor demand test [3]. It states that a multi-task system is schedulable if and only if the amount of *required* computation time is less than or equal to the amount of *available* computation time:

$$\Delta t \geq \sum_{\forall \tau_j} \left[\eta_j \left(\Delta t - D_j \right) * \left(c_j + o_j \right) \right] \tag{10.6}$$

According to the event-based task model described above, Δt denotes one time interval to be analyzed. $\eta_j(\Delta t - D_j)$ returns the number of activations of task τ_j that happen within Δt and that must be finished before the deadline D_j. Each task activation is multiplied by the task's respective maximum computational demand, i.e., its WCET plus additional preemption overheads o_j.

Since Eq. (10.6) is linear, it can directly be added to our multi-tasking ILP model for each task τ_j. This schedulability test has to be modeled for all possible time intervals Δt. The maximal interval to be considered is regularly given by the task set's hyperperiod. Checking all possible intervals Δt up to the hyperperiod is practically infeasible. Fortunately, task preemptions can only occur if a new task is ready for execution for many real-life scheduling policies like, e.g., EDF. Thus, the schedulability test from Eq. (10.6) has to be modeled in the ILP only at the points of discontinuity of the task set's density functions η. Finally, one constraint needs to be added that ensures that the system's overall load due to periodically repeating task activations stays below 100%. It is also possible to extend this approach towards fixed-priority scheduling, and the resulting ILP model grows only linearly with the number of events that have to be analyzed, in contrast to the quadratic nature inherent to [22, 24].

Figure 10.1 shows the effect of our ILP-based multi-task SPM allocation on schedulability of task sets featuring 8 tasks. We randomly selected 20 different task sets from TACLeBench [9]. Task periods were also randomly determined using UUniFast [4] and adjusted [39]. For each task set, periods were assembled such that the entire system has an approximate initial load of 0.8, 1.0, ..., 2.2 i.e., 8 different system loads are evaluated per task set. Task deadlines were chosen uniform randomly between 0.8 and 1.2 times the task's period. Furthermore, a jitter of up to 1% of each task's period was chosen uniform randomly. Our evaluation considered an ARM-based architecture with access latencies for main memory and SPM of 6 and 1 clock cycles, resp. The scratchpad size was set to 40% of each task set's total size.

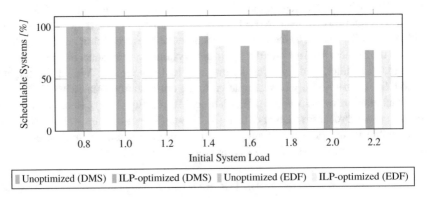

Fig. 10.1 Evaluation of schedulability-aware SPM allocation for 8 tasks

Figure 10.1 shows the schedulability of the task sets for the given initial system loads, using Deadline-Monotonic Scheduling (DMS) and Earliest Deadline First (EDF). The green and orange bars show the percentage of schedulable systems without any optimization applied, while the purple and yellow bars represent the schedulability after our ILP-based multi-task SPM allocation.

For an initial system load of 0.8, all task sets are schedulable, irrespective of the considered scheduling policy or whether the SPM allocation was applied or not. This is not surprising, since the considered systems feature sufficient idle times so that valid schedules are always found. The situation changes when considering higher initial system loads that range from 1.0 up to 2.2. In these scenarios, no task set was schedulable in an unoptimized state where the scratchpad memories were not used at all. However, our multi-task optimization is able to turn the vast majority of initially unschedulable task sets schedulable. For DMS scheduling, our ILP-based optimization achieves rates of schedulable task sets ranging from 100% (initial system loads of 1.0 and 1.2) to still 75% for an initial system load of 2.2. For EDF scheduling, the percentages of finally schedulable task sets are slightly smaller— they range from 95% (initial system load of 1.0) to 75% again. The time required to solve our ILPs is moderate. The whole compilation, analysis, and optimization process using a modern ILP solver like, e.g., `gurobi` required less than 6 CPU minutes on average over all considered task sets.

10.4 Analysis and Optimization of Multi-Processor Systems on Chip

To address PREDATOR Challenge #2 on analyzing and shaping the communication traffic for MPSoC architectures, it is important to understand when events happen in a multi-core architecture which potentially influences the cores' timing behavior. For this purpose, modular performance analyses use so-called request functions α which are very similar to the density function η from Sect. 10.3. In the context of MPSoCs, however, such functions characterize how often a processor core requests the shared bus of a multi-core architecture within a certain interval of time. Usually, such functions are provided at a very abstract level assuming execution models consisting of, e.g., the aforementioned acquisition, execution, and replication phases. For a precise analysis when each core attempts to access a shared hardware resource, it is, therefore, beneficial to extract request functions at the machine code level [14, 27].

For a precise and tight MPSoC performance analysis, both lower and upper bounds of resource requests are generated. Positions within the machine code executed on the different cores are identified where timing-relevant requests are generated, i.e., where shared hardware resources are accessed. Based on the code's Control Flow Graph (CFG), all possible sub-paths inside the code that feature these identified positions have to be considered. For this purpose, the well-known Implicit

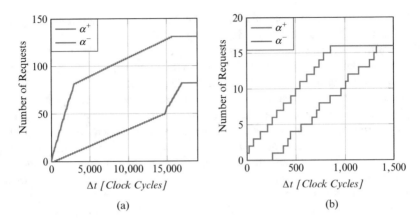

Fig. 10.2 Extracted request functions for selected benchmarks. (**a**) Compressdata. (**b**) binarysearch

Path Enumeration Technique (IPET) [17] has been modified to find the maximum number of requests potentially occurring in a given time interval along any path of a program. An algorithm has been proposed [27] that provides bounds on the number of requests for time intervals Δt of a program's runtime under consideration of all possible paths inside the CFG. This algorithm can be parameterized to trade precision of the generated request functions versus required execution time by varying the number of sampling points, i.e., the granularity of time intervals Δt considered by the algorithm.

Examples of lower (α^-) and upper (α^+) request functions generated for two selected benchmarks compressdata and binarysearch are shown in Fig. 10.2. The vertical distance between the lower and upper functions shows the variation of the number of produced requests. For example, compressdata can terminate with solely 82 shared bus accesses in total, or with up to 131. For binarysearch, both the lower and upper request functions converge to a common value, since each possible path through the program's code covers an identical number of bus requests. Only the points in time when these events occur differ.

Figure 10.3 shows the influence of the number of considered sampling points on the precision of the upper request function α^+ of compressdata. The finest-possible granularity, i.e., $\Delta t = 1$ clock cycle, leads to 131 samples in total and to a very smooth and precise result. When reducing the granularity such that only 50 samples are considered, the resulting request function has a clearly visible stepwise shape. However, the resulting function for 50 samples always dominates the most precise function so that no unsafe results are produced. For the highest precision with 131 samples, our algorithm requires 48 CPU seconds. In contrast, the time required to generate the request function for compressdata decreases down to 10 CPU seconds if 50 sampling points are considered.

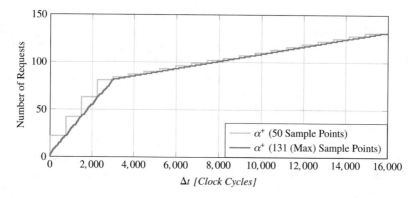

Fig. 10.3 Request functions for `compressdata` with different precision levels

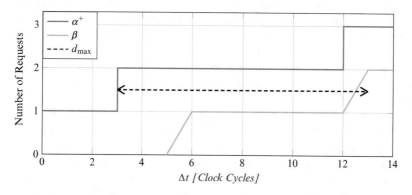

Fig. 10.4 Request functions α and delivery functions β

While request functions α denote the resource demand of a task w.r.t. shared bus accesses, so-called delivery functions β model the available capacity of a shared hardware resource during modular performance analysis [12, 35]. The relationship between both types of functions is illustrated in Fig. 10.4. The maximal horizontal distance between α^+ and β represents the maximum delay d_{max} a task exhibits due to blocked shared bus requests. In the figure, a task requests 2 bus accesses during interval lengths of 3 clock cycles. However, the bus can deliver the desired capacity only within 13 clock cycles. Thus, a blocking time of 10 clock cycles results from Fig. 10.4.

If a compiler could modify the generated code such that a task's request function is shifted towards the rightmost end of Fig. 10.4, its blocking time gets reduced which in turn probably decreases WCRTs and improves schedulability for the entire MPSoC system. This approach was investigated by a Master's Thesis [28] where instruction scheduling was exploited. Locally within basic blocks, those instructions requesting shared bus accesses were postponed by scheduling independent instructions in front of them. If this succeeds for all program paths of a given length

Δt (e.g., for $\Delta t = 3$ in Fig. 10.4), then the request functions are actually shifted as intended. This work revealed that compilers can be enabled to systematically reduce blocking times this way. For MPSoC task sets generated from the MRTC [11] and UTDSP [37] benchmark collections, blocking time reductions of up to 22.5% were reported. A solely local rescheduling of instructions, however, suffers from the inherent limitation that there is not too much potential for postponing shared bus accesses within a single basic block. Thus, maximal WCRT reductions of only up to 7.3% were achieved.

This basic idea to reshape bus requests at the code level is also pursued in currently ongoing work. By transforming the behavior of a task, its request function is modified such that its traffic will match a required profile. This is done by inserting additional machine instructions into the code, i.e., NOPs. Therefore, this approach does not rely on specific hardware or on operating systems that realize traffic shaping. Instead, the notion of code-inherent traffic shaping is introduced. If the places where to insert such additional instructions in a task's CFG are carefully chosen, parts of its request function that do not fit to a given access profile can be shaped systematically, even without necessarily increasing the task's WCET. For this purpose, two shaping algorithms using a greedy heuristic and an evolutionary algorithm have been designed which support various kinds of Leaky Bucket shapers [36].

The effectiveness of code-inherent shaping is depicted in Fig. 10.5 by means of MRTC's select benchmark. Based on a Leaky Bucket that generates a stepwise shaping profile, a delivery function β is assumed such that only half of the requests originally issued by the task within 1000 clock cycles can be fulfilled. It can be seen that the systematic insertion of a total of 408 NOP instructions results in a shaped request function that always stays below this delivery function. For this particular select task, its WCET increases from originally 36,019 clock cycles up to 50,317 clock cycles. While this WCET increase by 40% seems disadvantageous at a first glance, it is absolutely acceptable if the task still meets its deadline and if the shaped request function enables schedulability of the entire MPSoC task set.

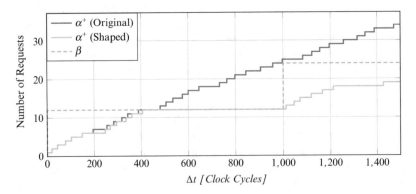

Fig. 10.5 Traffic shaping of select with $\beta(\Delta t)$ being 50% of $\alpha(\Delta t)$ for $\Delta t = 1000$

10.5   Multi-Objective Compiler Optimizations Under Real-Time Constraints

The simultaneous consideration of multiple optimization objectives by a compiler according to PREDATOR Challenge #3 can, to some extent, already be achieved using ILP-based techniques, even though ILPs only allow for one objective function to be maximized or minimized. PREDATOR's distinction between efficiency requirements on the one hand and worst-case constraints on the other hand naturally suggests to model critical constraints that must always be fulfilled as inequations in an ILP. Efficiency requirements are then modeled by an ILP's objective function and get optimized in addition to the satisfaction of critical constraints. This way, it is rather straightforward to turn the multi-task scratchpad memory allocation described in Sect. 10.3 into a multi-objective WCET-, schedulability- and energy-aware optimization.

The schedulability tests from Eq. (10.4) or (10.6) are mandatory constraints in the SPM allocation's ILP model. Using an energy model like, e.g., [32], the energy consumption e_i of each basic block b_i can be characterized in dependence of the ILP's binary decision variables x_i. By combining these block-level energy values with profiling-based information about the blocks' execution frequencies, the overall energy consumption e_j of task τ_j can be modeled. Multiplying these task-level energy values with the tasks' activation functions η_j (cf. Sect. 10.3) over the entire task set's hyperperiod H yields an expression that models the energy dissipation of the complete multi-task system and that thus can be minimized under simultaneous adherence to the ILP's schedulability constraints:

$$\min \sum_j \eta_j\,(H) * e_j \qquad (10.7)$$

Evaluation results for randomly generated sets of 6 tasks are depicted in Fig. 10.6, the experimental setup is the same as described in Sect. 10.3. Figure 10.6a shows the task sets' schedulability for their respective initial system loads, again using DMS and EDF scheduling. As can be seen, the multi-objective ILP is able to turn more than 95% of all task sets schedulable for initial system loads of up to 1.6. For higher initial loads, schedulability was still achieved for more than 70% of all task sets.

Simultaneously, considerable energy reductions compared to systems that do not use the SPM were achieved, cf. Fig. 10.6b. For initial system loads of up to 1.8, the task sets' energy dissipation was reduced down to less than 70%. For higher initial system loads, the resulting energy consumption still ranges from 71 to 77%.

Another common additional optimization goal is to meet code size requirements. Code compression might be used to meet code size constraints in embedded systems. However, the performance overhead of such techniques might be critical for real-time systems that must adhere to strict timing constraints. In the context of PREDATOR Challenge #3, we thus recently considered compiler-based code compression for hard real-time systems for the very first time [26]. This approach

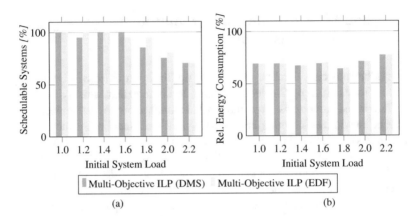

Fig. 10.6 Evaluation of multi-objective schedulability- and energy-aware SPM allocation for 6 tasks. (**a**) Schedulability. (**b**) Energy consumption

exploits lossless asymmetric compression algorithms [13] where a computationally demanding and highly effective code compression is performed at compile time, while the decompression is computationally lightweight so that it is feasible to perform it at runtime.

In the proposed approach, complete binary executable functions are selected and compressed by the WCC compiler and the resulting bit stream is added to the executable code produced by the compiler. Furthermore, the executable is extended by specifically tailored code for the decompression of the selected functions. Upon execution of a program optimized this way, all compressed functions are decompressed in one go during the program's start. For this purpose, a processor's scratchpad memory is used as a buffer that finally holds all decompressed functions. These functions are then directly executed from the SPM.

This approach trades code size reductions due to the selection of functions to be compressed with the decompression overheads in terms of WCET which should be as small as possible. For this purpose, an ILP is proposed whose binary decision variables x_i encode whether function f_i is compressed or not.

For each function f_i that might be compressed, its original, uncompressed code size S_i^{orig} and its Worst-Case Execution Time C_i^{orig} are pre-computed. Assuming that f_i would be compressed, the corresponding values S_i^{comp} and C_i^{comp} can also be pre-determined. For the WCET analysis of a potentially compressed function f_i, the decompression routine is added by the compiler, and the loops therein are precisely annotated with upper iteration bounds for the decompression of the currently considered function f_i in order to support the WCET analyzer aiT. Based on this data, the impact of f_i's compression on the entire program's code size ΔS_i and Worst-Case Execution Time ΔC_i can be expressed in the ILP.

ILP constraints ensure that the decompressed functions fit in the available SPM, that the entire program never gets larger due to the inserted decompression routine, and that the WCET increases of all functions always stay below a user-provided

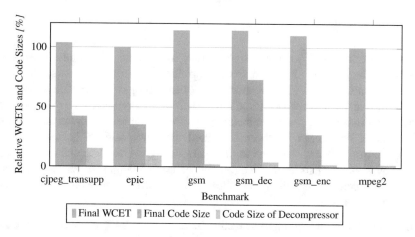

Fig. 10.7   Evaluation of compiler-based WCET-aware code compression for MediaBench

threshold ΔC^{limit}. Under these constraints, the ILP finally minimizes the entire program's code size by selecting appropriate functions f_i for compression.

For six large-sized benchmarks from MediaBench [16], the effects of the proposed compiler-based code compression for an Infineon TriCore architecture are depicted in Fig. 10.7. For each considered benchmark, the diagram shows the resulting relative WCETs and code sizes, as well as the code size of the decompression routine added by the compiler. The 100% baseline of Fig. 10.7 denotes the WCETs and code sizes of the original, unoptimized benchmarks, resp. For the ILP-based selection of functions to be compressed, the threshold ΔC^{limit} was set to 0.5 so that maximum WCET increases by 50% were still accepted by the optimization.

As can be seen from Fig. 10.7, the finally obtained WCET increases are way below this user-provided upper bound. For epic and mpeg2, the WCETs degrade only marginally by 0.6% and 0.5%, resp. The WCETs of the other benchmarks increase between 3.5% and 14.1% only. In contrast to this, our approach achieves rather large code size reductions. After the optimization of gsm_dec, its executable occupies only 73% of its original memory space. For all other benchmarks, an even higher degree of compression was achieved that reduces code sizes by more than a half. This way, the code size of cjpeg_transupp was reduced to 42% of its original size, and a maximal reduction down to only 13% of the original code size was achieved for mpeg2. Finally, Fig. 10.7 shows that adding extra code to the generated binaries for the decompression routine is worthwhile, since this overhead is over-compensated by the achieved overall code size reductions. As can be seen, the code size overhead due to the decompressor varies between 2% (gsm, gsm_enc and mpeg2) up to 15% (cjpeg_transupp) only, compared to the benchmarks' original code size.

10.6   Conclusions

This article presented a survey of work done in the field of compiler techniques for real-time systems in the authors' group during the past 10 years. Origin of all these activities was the collaborative research project PREDATOR funded by the European 7th Framework Programme. During this project, seminal work was carried out in order to design predictable yet efficient embedded systems. A couple of scientific challenges has been identified that have initially been considered during PREDATOR and that, due to their complexity, required continuous research effort over many years even after the end of this collaborative research project. This article summarized these compiler-centric activities and their corresponding scientific challenges:

Challenge #1:   Integration of task coordination into WCET-aware compilation
Challenge #2:   Analysis and optimization of Multi-Processor Systems on Chip
Challenge #3:   Predictable multi-objective compiler optimizations

Despite the advances in the field of compilation for real-time systems achieved in the past years, we expect that a continuation of this effort is necessary in the future. This is motivated by the trend towards massively parallel embedded real-time systems on the one hand, which still requires dedicated analyses and optimizations that are capable to support current and future many-core architectures. On the other hand, the simultaneous trade-off of various optimization objectives and the corresponding systematic exploration of the design space is still an unsolved problem for optimizing compilers. Last but not least, another important driver for future research is the increasing complexity of the involved system- and code-level analyses and optimizations which needs to be managed to obtain automated design tools that are usable in practice even for highly sophisticated and massively parallel systems.

Acknowledgments Parts of the work surveyed in this article received funding from Deutsche Forschungsgemeinschaft (DFG) under project No. 200265263 and 380772147. Other parts received funding from the European Union's 7th Framework Programme under grant agreement No. 216008 (PREDATOR) and from the Horizon 2020 research and innovation programme under grant agreement No. 779882 (TEAMPLAY).

References

1. AbsInt Angewandte Informatik GmbH, aiT: worst-case execution time analyzers (2020). http://www.absint.com/ait
2. K. Albers, F. Slomka, An event stream driven approximation for the analysis of real-time systems, in *Proceedings of the 16th Euromicro Conference on Real-Time Systems (ECRTS)* (2004). https://doi.org/10.1109/EMRTS.2004.1311020

3. S.K. Baruah, Dynamic- and static-priority scheduling of recurring real-time tasks. Real-Time Syst. **24**, 93–128 (2003). https://doi.org/10.1023/A:1021711220939

4. E. Bini, G.C. Buttazzo, Measuring the performance of schedulability tests. Real-Time Syst. **30**, 129–154 (2005). https://doi.org/10.1007/s11241-005-0507-9

5. European Commission, Grant Agreement for FP7-ICT-216008 PREDATOR (2007)

6. European Commission, Design for predictability and efficiency (2017). https://cordis.europa.eu/project/rcn/85432

7. H. Falk, J.C. Kleinsorge, Optimal static WCET-aware scratchpad allocation of program code, in *Proceedings of the 46th Design Automation Conference (DAC)* (2009). https://doi.org/10.1145/1629911.1630101

8. H. Falk, P. Lokuciejewski, A compiler framework for the reduction of worst-case execution times. Real-Time Syst. **46**, 251–300 (2010). https://doi.org/10.1007/s11241-010-9101-x

9. H. Falk, S. Altmeyer, P. Hellinckx, et al., TACLeBench: a benchmark collection to support worst-case execution time research, in *Proceedings of the 16th International Workshop on Worst-Case Execution Time Analysis (WCET)* (2016). https://doi.org/10.4230/OASIcs.WCET.2016.2

10. K. Gresser, An event model for deadline verification of hard real-time systems, in *Proceedings of the 5th Euromicro Workshop on Real-Time Systems (ECRTS)* (1993). https://doi.org/10.1109/EMWRT.1993.639067

11. J. Gustafsson, A. Betts, A. Ermedahl, B. Lisper, The Mälardalen WCET benchmarks: past, present and future, in *Proceedings of the 10th International Workshop on Worst-Case Execution Time Analysis (WCET)* (2010). https://doi.org/10.4230/OASIcs.WCET.2010.136

12. R. Henia, A. Hamann, M. Jersak, R. Racu, R. Richter, R. Ernst, System level performance analysis – the SymTA/S approach, in *IEE Proceedings – Computers and Digital Techniques* (2005). https://doi.org/10.1049/ip-cdt:20045088

13. A. Hidayat, FastLZ – free, open-source, portable real-time compression library (2007). http://fastlz.org

14. M. Jacobs, S. Hahn, S. Hack, WCET analysis for multi-core processors with shared buses and event-driven bus arbitration, in *Proceedings of the 23rd International Conference on Real-Time Networks and Systems (RTNS)* (2015). https://doi.org/10.1145/2834848.2834872

15. M. Joseph, P.K. Pandya, Finding response times in a real-time system. Comput. J. **29**, 390–395 (1986). https://doi.org/10.1093/comjnl/29.5.390

16. C. Lee, M. Potkonjak, W.H. Mangione-Smith, MediaBench: a tool for evaluating and synthesizing multimedia and communications systems, in *Proceedings of the 30th Annual International Symposium on Microarchitecture (1997)*. https://doi.org/10.1109/MICRO.1997.645830

17. Y.T.S. Li, S. Malik, Performance analysis of embedded software using implicit path enumeration, in *Proceedings of the Design Automation Conference (DAC)* (1995). https://doi.org/10.1145/217474.217570

18. C.L. Liu, J.W. Layland, Scheduling algorithms for multiprogramming in a hard-real-time environment. J. ACM (1973). https://doi.org/10.1145/321738.321743

19. P. Lokuciejewski, H. Falk, P. Marwedel, WCET-driven, code-size critical procedure cloning, in *Proceedings of the 11th International Workshop on Software and Compilers for Embedded Systems (SCOPES), Munich* (2008), pp. 21–30

20. P. Lokuciejewski, S. Plazar, H. Falk, P. Marwedel, L. Thiele, Multi-objective exploration of compiler optimizations for real-time systems, in *Proceedings of the 13th International Symposium on Object/Component/Service-oriented Real-time Distributed Computing (ISORC)* (2010). https://doi.org/10.1109/ISORC.2010.15

21. P. Lokuciejewski, S. Plazar, H. Falk, P. Marwedel, L. Thiele, Approximating Pareto optimal compiler optimization sequences – a trade-off between WCET, ACET and code size. Softw. Pract. Exp. (2011). https://doi.org/10.1002/spe.1079

22. A. Luppold, H. Falk, Code optimization of periodic preemptive hard real-time multitasking systems, in *Proceedings of the 18th International Symposium on Real-Time Distributed Computing (ISORC)* (2015). https://doi.org/10.1109/ISORC.2015.8

23. A. Luppold, H. Falk, Schedulability aware WCET-optimization of periodic preemptive hard real-time multitasking systems, in *Proceedings of the 18th International Workshop on Software & Compilers for Embedded Systems (SCOPES)* (2015). https://doi.org/10.1145/2764967.2771930

24. A. Luppold, H. Falk, Schedulability-aware SPM allocation for preemptive hard real-time systems with arbitrary activation patterns, in *Proceedings of Design, Automation and Test in Europe (DATE)* (2017). https://doi.org/10.23919/DATE.2017.7927149

25. A. Luppold, D. Oehlert, H. Falk, Evaluating the performance of solvers for integer-linear programming. Tech. Rep., Hamburg University of Technology (2018). https://doi.org/10.15480/882.1839

26. K. Muts, A. Luppold, H. Falk, Compiler-based code compression for hard real-time systems, in *Proceedings of the 22nd International Workshop on Software and Compilers for Embedded Systems (SCOPES)* (2019). https://doi.org/10.1145/3323439.3323976

27. D. Oehlert, S. Saidi, H. Falk, Compiler-based extraction of event arrival functions for real-time systems analysis, in *Proceedings of the 30th Euromicro Conference on Real-Time Systems (ECRTS)* (2018). https://doi.org/10.4230/LIPIcs.ECRTS.2018.4

28. N. Piontek, Instruktionsscheduling für harte Multi-Core Echtzeitsysteme mit gemeinsam genutztem Datenbus. Masters Thesis, Hamburg University of Technology (TUHH) (2018)

29. S. Plazar, P. Lokuciejewski, P. Marwedel, WCET-aware software based cache partitioning for multi-task real-time systems, in *Proceedings of the 9th International Workshop on Worst-Case Execution Time Analysis (WCET)* (2009). https://doi.org/10.4230/OASIcs.WCET.2009.2286

30. A. Schranzhofer, R. Pellizzoni, J.J. Chen, L. Thiele, M. Caccamo, Worst-case response time analysis of resource access models in multi-core systems, in *Proceedings of the Design Automation Conference (DAC)* (2010). https://doi.org/10.1145/1837274.1837359

31. A. Schranzhofer, J.J. Chen, L. Thiele, Timing analysis for TDMA arbitration in resource sharing systems, in *Proceedings of 16th IEEE Real-Time and Embedded Technology and Applications Symposium (RTAS)* (2010). https://doi.org/10.1109/RTAS.2010.24

32. S. Steinke, M. Knauer, L. Wehmeyer, P. Marwedel, An accurate and fine grain instruction-level energy model supporting software optimizations, in *Proceedings of the International Workshop on Power And Timing Modeling, Optimization and Simulation (PATMOS). Yverdon-Les-Bains* (2001)

33. V. Suhendra, T. Mitra, A. Roychoudhury, T. Chen, WCET centric data allocation to scratchpad memory, in *Proceedings of the 26th IEEE Real-time Systems Symposium (RTSS)* (2005). https://doi.org/10.1109/RTSS.2005.45

34. The PREDATOR Consortium, PREDATOR – design for predictability and efficiency (2011). https://www.predator-project.eu

35. L. Thiele, S. Chakraborty, M. Naedele, Real-time calculus for scheduling hard real-time systems, in *The 2000 IEEE International Symposium on Circuits and Systems. Proceedings. ISCAS 2000 Geneva*, vol. 4 (2000), pp. 101–104

36. J.S. Turner, New directions in communications (or which way to the information age?). IEEE Commun. Mag. (1986). https://doi.org/10.1109/MCOM.1986.1092946

37. UTDSP Benchmark Suite (2019). http://www.eecg.toronto.edu/$\sim$corinna/DSP/infrastructure/UTDSP.html

38. R. Wilhelm, D. Grund, J. Reineke, M. Schlickling, M. Pister, C. Ferdinand, Memory hierarchies, pipelines, and buses for future architectures in time-critical embedded systems. Trans. Comput.-Aid. Des. Integr. Circuits Syst. **28**, 966–978 (2009). https://doi.org/10.1109/TCAD.2009.2013287
39. J. Xu, A method for adjusting the periods of periodic processes to reduce the least common multiple of the period lengths in real-time embedded systems, in *Proceedings of the International Conference on Mechatronic and Embedded Systems and Applications (MESA)* (2010). https://doi.org/10.1109/MESA.2010.5552058

Index

© The Author(s) 2021
J.-J. Chen (ed.), *A Journey of Embedded and Cyber-Physical Systems*,
https://doi.org/10.1007/978-3-030-47487-4

Printed in the United States
by Baker & Taylor Publisher Services